Learning from the Asian Tigers

Studies in Technology and Industrial Policy

Sanjaya Lall
Lecturer in Development Economics
Queen Elizabeth House
University of Oxford

First published in Great Britain 1996 by
MACMILLAN PRESS LTD
Houndmills, Basingstoke, Hampshire RG21 6XS
and London
Companies and representatives
throughout the world

A catalogue record for this book is available
from the British Library.

ISBN 0–333–67410–3 hardback
ISBN 0–333–67411–1 paperback

First published in the United States of America 1996 by
ST. MARTIN'S PRESS, INC.,
Scholarly and Reference Division,
175 Fifth Avenue,
New York, N.Y. 10010

ISBN 0–312–16560–9

Library of Congress Cataloging-in-Publication Data
Lall, Sanjaya.
Learning from the Asian tigers : studies in technology and
industrial policy / Sanjaya Lall.
p.   cm.
Includes bibliographical references and index.
ISBN 0–312–16560–9 (cloth)
1. Industrial policy—Asia.   2. Asia—Economic conditions—1945–
I. Title.
HD3616.A773L35   1997
338.95—dc20
96–34348
CIP

10   9   8   7   6   5   4   3   2   1
05   04   03   02   01   00   99   98   97   96

Printed in Great Britain by
The Ipswich Book Company Ltd
Ipswich, Suffolk

LEARNING FROM THE ASIAN TIGERS

This book is dedicated to our son
Ranjit Lall

Contents

List of Tables

List of Figures

Preface and Acknowledgements

Technology development and industrial policy continue to be important areas of research and policy interest in development economics. The launching of liberalisation and adjustment policies in the developing world makes them of even greater relevance, since the success of policy reform depends crucially upon the ability of industry to bring technical efficiency to world standards and use the limited resources available to best effect. This collection of papers brings together some of my recent work, much of it published in academic journals, on these issues. It builds upon detailed empirical work on the process of technological learning in enterprises in developing countries. It also draws upon advisory work that I have done recently for various governments and international institutions in Asia and Africa. Much of this detailed work cannot be reported here, but the accretion of knowledge and experience has fed into the general analysis.

Several of the papers challenge the ruling orthodoxy on the subject of industrial policy – the roles of markets versus governments in the process of industrial development. They argue that the current neoliberal consensus on the virtues of free markets overplays the magic of the marketplace. Accepting that much of the past pattern of government intervention was misguided and costly, the research reported here suggests that the existence of pervasive market failures in the developing world calls for remedial and strategic intervention by governments. The evidence of the most successful industrialising countries in the world, the NIEs of East Asia, supports the case for government intervention, albeit in a very different form from that practised earlier. Their experience also shows that there are many different modes of intervention possible: the Asian experience spans a wide range of approaches and results. Some interventions were 'market friendly', others were highly selective and targeted. Both coexisted with free trade and protection. Foreign direct investment played very different roles in the different NIEs. The levels of industrial depth and technological development achieved were also different. An understanding of these differences, and the underlying policy factors, can itself constitute a valuable contribution to the art of development policy.

This book is aimed at students of development as well as at practitioners. It covers the theoretical issues of technology and industrial policy, and reviews in some detail the actual policies adopted. It analyses the success stories as well as the failures, using the 'technological capability' approach that I have worked on for the past decade or so. It illustrates the utility of looking at broad strategic issues from the microeconomic perspective of a developing-country enterprise struggling to become competitive in a setting of deficient markets and rapidly changing technologies.

Most of the chapters have been published before, though I have made some minor changes for the purposes of this book. I wish to acknowledge the kind permission given by various journals to reprint. The sources of each of the chapters are as follows:

1. 'Paradigms of Development: the East Asian debate', *Oxford Development Studies*, vol. 24, no. 2 (April 1996).
2. 'Understanding Technology Development', *Development and Change*, vol. 24, no. 4 (October 1993), pp. 719–53.
3. 'Technology Development Policies: Lessons from Asia'. Not previously published.
4. '"The East Asian Miracle" Study: Does the bell toll for industrial strategy?', *World Development*, vol. 22, no. 4 (April 1994), pp. 645–54.
5. 'Structural Adjustment and African Industry', *World Development*, vol. 23, no. 12 (December 1994), pp. 2019–31.
6. 'Malaysia: Industrial success and the role of government', *Journal of International Development*, vol. 7, no. 5 (September 1995), pp. 759–73.
7. 'Skills and Capabilities in Ghana's Competitiveness' (with Ganeshan Wignaraja). Not previously published.
8. 'Foreign Direct Investment Policies in the Asian NIEs', published as 'Developing Asian Economies' in J. H. Dunning (ed.), *Governments, Globalization and International Business* (Oxford: Oxford University Press, 1996).

List of Abbreviations

ASEAN	Association of South East Asian Nations
CAD	Computer-aided design
CAE	Computer-aided engineering
CAM	Computer-aided manufacture
CETRA	China External Trade Development Council
CEO	Chief executive officer
CPC	China Productivity Centre
CSIR	Council of Scientific and Industrial Research
DRAM	Dynamic random access memory
EDB	Economic Development Board
E-O	Export-oriented
ERSO	Electronics Research and Service Organisation
FDI	Foreign direct investment
FTZ	Free trade zone
GATT	General Agreement on Tariffs and Trade
GDP	Gross domestic product
GNP	Gross national product
HAN	Highly Advanced National Project
HCI	Heavy and chemical industries
HICOM	Heavy Industry Corporation of Malaysia
HKDIC	Hong Kong Design Innovation Company
HKPC	Hong Kong Productivity Council
H-O	Heckscher-Ohlin
HPAE	High performing Asian economy
ICICI	Industrial Credit and Investment Corporation of India
IDB	Industrial Development Bank
III	Institute for the Information Industry
IMCB	Institute for Molecular and Cell Biology
IMF	International Monetary Fund
IMP	Industrial Master Plan
I-S	Import-substitution
ISO	International Standards Organisation
ITAF	Industrial Technical Assistance Fund
ITD	Industrial technology development
ITRI	Industrial Technology Research Institute
KAIS	Korea Advanced Institute of Science

KAIST	Korea Advanced Institute of Science and Technology
KDB	Korea Development Bank
KIST	Korea Institute of Science and Technology
KIT	Korea Institute of Technology
KOTRA	Korea Overseas Trade Agency
KTAC	Korea Technology Advancement Corporation
KTCGF	Korea Technology Credit Guarantee Fund
KTDC	Korea Technology Development Corporation
LIUP	Local Industries Upgrading Programme
MIDA	Malaysian Industrial Development Authority
MNC	Multinational corporation
MSTQ	Metrology, standards, testing and quality
MTDC	Malaysia Technology Development Corporation
MVA	Manufacturing value added
NBER	National Bureau of Economic Research
NDP	New Development Policy
NEP	New Economic Policy
NIE	Newly industrialising economy
NTB	National Technology Board
NUS	National University of Singapore
OECD	Organisation for Economic Cooperation and Development
OED	Operations Evaluation Department
OEM	Original equipment manufacture
OTA	Office of Technology Assessment
PCB	Printed circuit board
PSDC	Penang Skills Development Centre
QC	Quality control
R&D	Research and development
RAM	Random access memory
SAP	Structural adjustment programme
SBI	Singapore Bio-Innovation
SDF	Skills Development Fund
SISIR	Singapore Institute of Standards and Industrial Research
SITAS	Small Industries Technical Assistance Scheme
SITC	Standard Industrial Trade Classification
SME	Small and medium enterprise
SMI	Small and medium industry
SPREAD	Sponsored research and development programme
S&T	Science and technology
TC	Technological capability
TDC	Technology Development Centre

TDR	Technology Development Reserve
TFP	Total factor productivity
TSMC	Taiwan Semiconductor Manufacturing Company
UMC	United Microelectronics Company
UNDP	United Nations Development Programme
UNIDO	United Nations Industrial Development Organisation
VC	Venture capital
VDP	Vendor Development Programme
VITB	Vocational and Industrial Training Board
VLSI	Very large-scale integrated (circuits)
WTO	World Trade Organisation

1 Paradigms of Development: The East Asian Debate on Industrial Policy[1]

INTRODUCTION

The debate on economic 'paradigms' addressed here concerns industrial policy: the role of governments in promoting industrial development beyond what would be possible without intervention. The countries that this debate concerns in East Asia are the four NIEs – Korea, Taiwan, Singapore and Hong Kong. These countries are among the most successful industrialisers in the developing world. Much of the debate on industrial policy has revolved around their experience.

The growing dominance of neoclassical development economics has centred on one paradigm of development and industrialisation policy. Despite some variations, the paradigmatic element comprises a set of common basic themes: markets are basically efficient and governments basically inefficient, resource allocation is optimised by agents responding to free markets, and the *best development policy is to remove all interventions in the functioning of free markets*. This paradigm combines simple neoclassical theory with certain empirical assumptions about how economies and governments function, what drives growth and structural transformation, and what 'good' development policy consists of. Neoclassical theory *per se* does not lead to this development paradigm, since a different set of empirical premises could lead to very different policy conclusions with the same economic apparatus; it is important therefore to distinguish theory from development economics. It is the latter that now informs much of policy making and has the powerful backing of most aid donors and leading international institutions. It constitutes the framework for the emerging 'rules of the game' for economic life, and it is important that it should be rigorously examined.

There is increasing concern about some of the underlying assumptions and values of the new paradigm: about market efficiency, government inefficiency, the links from static optimisation to dynamic

1

growth, and the role of government interventions in explaining the recent industrial success. Were free markets really 'efficient' in the textbook sense, or at least the most efficient form of organising activity available to developing countries? Why did export orientation in the Asian NIEs turn out to be an efficient strategy – because it conformed to theoretical precepts of static optimisation, or because it introduced a different set of dynamic incentive effects and externalities than considered by standard neoclassical explanations? Will sweeping liberalisation in other countries really lead them into becoming NIEs in the Asian mould?

This chapter cannot go into all these issues, but focuses on the case for government intervention in industrial development. It starts with the evolution of the current debate about particular forms of intervention. It reviews the economic case for government intervention and the experience of the NIEs, and it closes with some lessons.

THE EVOLUTION OF THE DEBATE

The rise of the neoclassical development paradigm can be traced to many political and economic factors. Among the most important, at least as far as the developing world is concerned, was the economic success of the Asian NIEs. Over the 1970s evidence accumulated on the superior economic and industrial performance of some East Asian countries, with their private-sector-led, export-oriented industrialisation, over the state-led, highly planned and pervasively regulated import-substituting models elsewhere. This coincided with the rise of neo-liberalism in the USA and UK and cumulated into a case, not merely for export orientation, but generally for market-driven, non-interventionist policies in developing countries.

Before this, most development economists, disillusioned with the lack of market-driven development in the colonial' period, believed that developing-country markets were missing or inefficient. Widespread externalities were accompanied by severe informational, institutional and entrepreneurial gaps. The exploitation of externalities, scale economies and the coordination of investment decisions necessitated planning. Many accepted a large role for the public sector, in the provision of 'public goods' as well as in directly productive activities. The catalysing of domestic entrepreneurship called for strong signals. Since fledgling industrial activities in developing countries were considered unable to compete in world markets, the case for protecting new industries was accepted to give a learning base at home. In addition, the domestic market was taken to give advantages of lower transactions costs for new enterprises in developing countries.

This 'paradigm' noted (correctly) many of the market failures affecting development and the critical role for the state in guiding and coordinating industrial and entrepreneurial activity. The policy implications drawn were not, however, based on a coherent analysis of the nature of market failures, and overlooked the information and capabilities needed by governments to remedy them. In the initial enthusiasm for planned development, industrialisation was promoted mostly by import-substitution strategies, often in the context of comprehensive planning. Import-substitution offered across-the-board, perpetual protection for all activities, accompanied by a battery of other interventions unrelated to market failures. It disregarded considerations of efficiency, scale and specialisation, simply assuming that all the industries set up would be efficient, and ignoring the need for competitive signals to create and maintain efficiency. It was not selective in promoting new activities or gearing improvements in factor markets to the activities being fostered. It thus gave the wrong signals for resource allocation, and did not effectively exploit the beneficial externalities and linkages that some ('strategic') industries could offer. Moreover, by ignoring the extremely limited investment and human resources available to create efficient industries, it spread available resources too thinly. It thus turned out to be a blunt and rather inefficient tool of industrial promotion.

While this manifestation of the early paradigm was not a necessary consequence of the theory, the evidence on East Asian success was taken to contradict all its premises. It was not just trade orientation that was at issue. Export orientation became identified with 'neutrality' in trade, which in turn became equated with free trade, openness to all other forms of foreign transactions (in direct investment and technology flows), 'neutrality' in domestic resource allocation, and finally with liberal ('minimalist') governments that provided basic public goods, a legal framework and the rules of the game, and managed the macroeconomy well.[2] This transition in the development literature from export orientation to neoliberal political economy was smooth and imperceptible, and at the time persuasive. The theory on which economists were brought up was apparently alive and kicking in the developing world. The evidence of the NIEs suggested that standard H-O models of comparative advantage were borne out in practice, fears about market failure were unfounded, and countries that intervened strongly in markets suffered from gross inefficiencies.

The new paradigm thus emerged with a strong consensus. There was an implicit acceptance that simple textbook models described adequately the reality of the developing world: markets were fully efficient for both products and factors, the institutions needed for the functioning of markets were present or would arise naturally in response to market signals, there were no externalities or other distortions that needed to be addressed. There were simple, attractive and universal solutions to all development problems. No one may have believed that all the theoretical requirements of efficient markets (i.e. perfect competition) were met in developing countries. However, it was probably believed that existing markets were *efficient enough* to achieve the optimality predicted by theory (these issues were not discussed explicitly in the context of industrial policy). Thus, externalities were of limited practical significance, price signals led to efficient allocation in both product and factor markets, and, for present purposes a critical assumption, industries started and operated in a neoclassical world (selecting, mastering, using and improving technologies costlessly and risklessly) with no need for support or intervention to become efficient.

Plausible as this was, it faced a theoretical problem. In traditional neoclassical growth theory, static optimisation by itself could not account for the sustained high rates of growth seen in East Asia.[3] These models assumed diminishing returns to investment and predicted a convergence of growth rates across countries. Higher growth over time in particular regions could only be explained (given neoclassical assumptions) by introducing productive factors that enjoyed increasing returns and were endogenous to the economic system rather than being exogenously given. In the late 1980s 'new' growth theory provided such factors, essentially human capital and technology. These factors also entailed significant externalities, and private agents would underinvest in their creation in relation to social needs. *Ergo* there was a case for government intervention to remedy the market failures caused by the divergence between private and social benefits. Endogenous growth theory did not immediately affect the neoclassical approach to East Asia, but over time it proved valuable to the debate on industrial policy.

By the late 1980s irrefutable evidence had accumulated that most NIEs did not conform to the neoliberal characterisation.[4] They were aggressively picking or creating 'winners' at the industry (and even firm) level by intervening in trade, credit allocation, technology imports and local technology diffusion and creation, education and training, export activity and so on. The results were unprecedented rates of growth and diversification of manufacturing industry and exports, though with marked differences among the countries reflecting their differing levels and kinds of intervention (below). This presented a dilemma – either the interventions were desirable and there were pervasive market failures (in which case the neoclassical development paradigm was undermined), or the interventions were irrelevant despite being pervasive (in which case explanations were needed as to why they were undertaken, what they achieved, and why they did not lead to the kind of inefficiencies associated with them elsewhere).

The challenge was taken up by the World Bank in *The East Asian Miracle*. The effort was partly in response to a controversial internal study (in which I participated) by its Operations Evaluation Department that criticised the biased interpretation by the Bank of its own evidence on Korea (OED, 1992), and partly in response to demands by the Japanese government for less ideological policy advice. The study drew a distinction between desirable 'market-friendly' and other, undesirable, interventions. Market-friendly interventions were defined as 'functional' – those that did not try to direct resources to particular activities, but remedied generic failures in markets. Non-market-friendly ones were 'selective', influencing resource allocation in favour of 'winners' picked by the government. The *Miracle* study explained the success of East Asia with reference to market-friendly interventions, arguing that selective interventions, while present, were unnecessary and contributed little, if anything, to East Asian success. There were no reasons in theory for selectivity, and no benefits to other countries from adopting these market-unfriendly policies, especially because they lacked the unique political economy to administer such policies.

The *Miracle* study was an important step forward in the industrial policy debate. It departed from the earlier Bank approach by admitting that *some* markets actually did not function efficiently, and that government intervention was needed to remedy market failure. It also admitted the existence and pervasiveness of selective interventions in East Asia. However, it was obliged to defend the fundamental postulates of the World Bank's policy advice – that governments should not be selective in influencing resource allocation, and, in particular, should not mount industrial policy. Thus, it redrew the lines of the legitimate functions of the government around a 'market-friendly' set of policies which were confined to support for human-capital formation (health and education), openness to information flows (technology inflows from abroad), and export promotion (export activity was believed to create generic externalities). The explanation for rapid growth based on functional policies drew support from 'new' growth theory, and it was assumed that all human-capital interventions were non-selective in their nature (note that new growth theory says little about the selectivity or otherwise of the interventions needed to promote human capital).

In the meantime, a growing literature on capability building in developing countries approached industrial policy from a different vantage point, that of micro-level technical change.[5] Drawing upon the 'evolutionary' approach (Nelson and Winter, 1982), it drew upon a wide base of empirical research, and focused on market failures affecting the development of capabilities at the enterprise level. In spirit this harked back to the earlier structuralist paradigm of policy making, though it drew upon neoclassical analysis. It married this analysis with the conduct of industrial policy in East Asia, and drew very different conclusions from that of the *Miracle* study on the significance of government interventions.

This is where the industrial policy debate in Asia rests at the moment. The neoclassical orthodoxy is in an uncomfortable limbo between strict neoliberal principles and the market-friendly half-way house proposed by the World Bank. The Bank has been severely criticised for its approach,[6] but has found itself unable to refute its critics. The debate itself is being overtaken by the wave of liberalisation and globalisation that is sweeping the world in the wake of adjustment programmes, the WTO, spreading 'globalisation' and so on. But this does not mean that the underlying case for intervention has been proved wrong.[7] Let us look therefore at this case.

THE CASE FOR INDUSTRIAL INTERVENTION

The World Bank's analysis of market-friendly versus selective interventions is perhaps a good place to start, since this is the last word on industrial policy by the leading exponent of development policy. There are two broad issues at stake. Is the distinction valid? And are selective interventions really unjustified in theory?

On the *first*, there are no theoretical grounds for distinguishing between 'market-friendly' functional and selective interventions: *any* policy that remedies market failure is 'friendly' to the market. On empirical grounds it may be argued that functional interventions have greater chance of success: governments can never exercise selectivity effectively, whereas they can find and remedy functional failures successfully. This may or may not be true. The evidence of East Asia suggests that selectivity was effectively used, and that not remedying the kinds of market failures that call for selective interventions can lead to stunted industrial development.

The particular definition of market-friendly interventions used by the Bank is also suspect. Are all the interventions suggested by new growth theories – skill formation or openness to technology inflows – always non-selective? Not necessarily. The creation of skills at the school level, and in some tertiary education, are broadly non-selective. However, certain vocational training, university-level technical and scientific education, and specialised industry training (either pre- or post-employment), can be extremely selective. If the pattern of investment in skills is closely geared to industrial promotion by protection or credit direction, then the former becomes just as selective as the latter. The East Asian evidence suggests that many education and technology import policies were in fact extremely selective, with close government direction of the content of enrolments and curricula to ensure conformance with the thrust of industrial policy.[8]

The *second* issue is more important: the case for selective intervention. In theory, according to the *Miracle* study, there may be four market failures in resource allocation:

- capital-market deficiencies (caused by information gaps),
- lumpiness of investment (scale economies),
- imperfect appropriability of firm-level investments in knowledge and skills, and
- the inability of individual investors to act rationally when there are technologically interdependent investments.[9]

The restoration of efficient resource allocation calls for intervention to coordinate investments and counteract externalities. The intervention may not be functional, since different activities, with differing technological characteristics and spillovers, may suffer to different degrees from these failures. The *Miracle* study, having noted the theoretical case, makes no attempt to analyse its empirical significance in East Asian industrialisation, or to find out whether or not governments attempted to remedy the failures, and, where they did, how successful they were.

While these four market failures provide valid arguments for selective intervention, they are not all the failures that affect industrial development, nor even the most important ones. They are derived from a simplified set of assumptions that essentially ignores the slow, costly, risky and largely unpredictable process by which firms in developing countries become efficient. *Industrial development faces important market failures, which provide some of the most critical arguments for selective intervention.*

The neoclassical depiction of industrial development assumes that technology is freely available from a known 'shelf', from which firms choose according to their factor and product prices. This technology is then absorbed costlessly and risklessly and used at 'best-practice' levels.[10] There is no need for intervention to support the process, and by definition any tampering can only lead to inefficiency in the choice and use of technology. There is an even stronger premise: any actual inefficiency *must* be due to interventions in efficient markets, and the removal of such interventions will be necessary and sufficient for restoring efficiency. Only 'good' and 'bad' firms exist, and they can only be sorted out by free markets.

If there is any lag in efficiency it can, at most, only be for a brief period in which scale economies are fully realised or costs fall in a 'learning by doing' process. However, these are taken to be predictable (scale economies are given by technical-design parameters, while the learning curve is taken to be known) and a simple function of the quantity of output: there is no need for intervention because firms can anticipate the process perfectly and raise money in efficient capital markets to finance the process. If there is failure in capital markets, the theoretical solution is to improve their functioning rather than to intervene selectively to support particular activities. Thus, capital-market failures and scale economies may not provide grounds for selectivity unless these failures cannot be remedied, and protection or subsidies are used as second-best solutions.

The capabilities literature suggests that this is oversimplified and misleading. Technology has many 'tacit' elements and cannot be transferred like a physical product. Its mastery and use require the recipient to invest in new skills, technical information, organisational methods and external linkages. The process continues over time, and varies by technology. It may be relatively short, cheap and predictable in 'easy' technologies where the knowledge is more embodied in simple equipment, the range of skills is limited, and the operation is relatively self-contained in an enterprise. In technologies that have complex processes and sophisticated equipment, the range of skills is large, there are many differing stages of production and large numbers of operations have to interact in the value-added chain; mastery may be prolonged, costly and risky. When firms are undergoing learning, it is difficult to sort out 'good' and 'bad' firms, since there is a large intermediate category.

More important, the process of learning in developing countries may be distorted and curtailed if firms do not *know* how to go about learning, how long it will take, how much it will cost, or where to look for information and skills. There may be a 'learning to learn' process (Stiglitz, 1987), which firms facing full international competition may be unwilling to undertake. Dropping the assumptions on perfect information of technology markets and transferability of technology (with no tacit elements or learning periods) thus poses market failures in resource allocation. Given the cost, risk and information gaps within the firm in learning, in free markets firms will tend to underinvest in technologies that have costly, prolonged and risky learning periods. This will also affect the process of technological deepening: entering more complex technologies, increasing local content, or undertaking more demanding technological tasks (say, from simple final-assembly technology to design and development activity).

The capability approach does not suggest that *no* industry will take root in free markets. Where there is a modicum of skills, infrastructure and low labour costs, simple labour-intensive activities may start (though in modern industry even the simplest of industries require advanced technical and management skills). However, entry into more complex and demanding technologies may be limited by the absence of supportive interventions to overcome learning costs. Such interventions cannot be functional – since technologies differ in their learning needs, they *have* to be selective.

The protection of infant industries is one, and historically the most popular and effective, means of remedying the failures.[11] However, protection is a dangerous tool. Apart from the cost to the consumer, it dilutes the incentive to invest in capability development, the very process it is meant to foster. Firms are very sensitive to competitive pressures in deciding to invest in capabilities, and the protection offered in typical I-S regimes has tended to detract from costly and lengthy investments in competitive skills and knowledge. There may be many solutions: offer limited protection (the Mill proposal); impose performance requirements; or enforce early entry into export markets while maintaining domestic protection. The last has the advantage that it exploits the externalities generated by export activity, and was the one used widely by the larger NIEs (that developed the deepest and most diverse industrial sectors).

Since firms do not learn on their own, however, protection can only *partly* remedy market failure. Firms draw upon a number of other firms and markets for capability development: input and equipment suppliers,

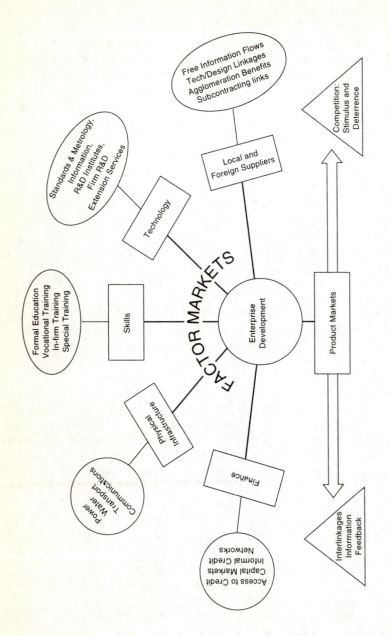

Figure 1.1 Enterprises and markets

skills, finance, technology, market information, and infrastructure. Figure 1.1 shows schematically the different markets in which a firm operates. All these markets may suffer from deficiencies. *Offering protection without remedying these can be wasteful, while simply improving factor markets without offsetting market failures to learning within firms can lead to narrow and shallow technological development.*

Most of the market failures outside firms are well recognised. The coordination problem caused by technological linkages between firms are noted in the *Miracle* study. The case for intervention, however, goes beyond simply coordinating individual investment decisions when there are externalities. The nature of the externalities is also important – when certain industry 'clusters' generate strong benefits for the economy in terms of technological learning, spillovers and dynamism, there may be a case for promoting them over others that have more limited or static effects. This case for 'strategic' sectors is noted by some new growth theorists, who distinguish between specialisations that lead cumulatively to technological stagnation or dynamism (Young, 1991). Arrow (1962) noted the risk of underinvestment in skills and technology because of inappropriability and leakages. Failures in markets for finance, skills, technology and infrastructure are universally accepted. All countries, developed and developing, have undertaken measures to remedy them, often selectively. These measures may involve creating new markets and institutions, or they may involve encouraging large firm size to enable the internalisation of the deficient markets (both were practised in East Asia).

Market failures are particularly binding for *local* enterprises, and even more so for new small and medium-sized entrants into modern industry (SMEs). Foreign investors, especially affiliates of large MNCs, face fewer failures in developing countries. Their *raison d'être* lies in the internalisation of many intermediate markets, especially for capital, skills and technology. This is why MNCs can be a powerful means of launching industrialisation in developing countries (as long as some complementary factors exist). Their significance is rising in activities where technologies are changing rapidly, production is growing more linked across nations, and export market access is growing more difficult for new entrants. However, the advantage offered by FDI does not mean (as neoclassical theory suggests) that the best way to develop is to adopt passive open-door policies in concert with free trade and other non-interventionist policies. There may be important market failures in the FDI process that call for interventions with the entry process and factor markets that affect MNC activity.[12]

First, a passive liberal policy may only attract MNCs into areas of static comparative advantage. Selective and functional interventions can guide FDI into dynamic and more complex activities (Singapore's strategy). Secondly, MNCs tend to transfer operating know-how rather than complex technological functions to developing host economies. The design and development process remains in advanced countries near sophisticated suppliers, R&D systems and skills. However, as countries industrialise it becomes increasingly important to develop R&D capabilities, to keep abreast of and absorb technologies, deepen industry and reduce the cost of importing technology. Again, there is a need to intervene to induce an upgrading of MNC technological activity (as in Singapore), or to restrict foreign entry as local firms have to establish their own innovative base. The latter strategy is designed to develop *indigenous* R&D capabilities, to capture the greater externalities and dynamic benefits that this may offer (as in Korea and Japan).

Theory thus provides valid grounds for interventions to promote industrial development. Market failures can take three forms: *within firms*, *in inter-firm relations* and *in factor markets* (Figure 1.2). These failures are inter-related, and their remedy calls for a range of *selective and functional* interventions. Those within firms have to be dealt with by providing a 'cushion' for learning (e.g. by protection), and by the provision of information and other support; those between firms, by the coordination of investments (partly by protection), geographical clustering and promotion of linkages; and those in factor markets, by direct interventions to remedy the failure. Note that protection meets only a small part of the need (within firms and in inter-firm relations); used by itself, it can be harmful for technological development because it leaves other failures untouched.

The TC (technological capabilities) approach is an advance in the industrial policy debate, offering a comprehensive and coherent view of policy needs. It explains why I-S strategies failed and E-O worked, not by 'getting prices right' and realising static comparative advantage, but by promoting a healthy and dynamic learning process. To conclude with some generalisations:

- Interventions in factor and product markets have to be closely coordinated and integrated; one without the other may be ineffective, even counter-productive. Factor-market policies as recommended by new growth theories cannot provide a complete explanation of rapid industrial development by indigenous enterprises, since they ignore the costs of learning and the variety of market failures faced.

13

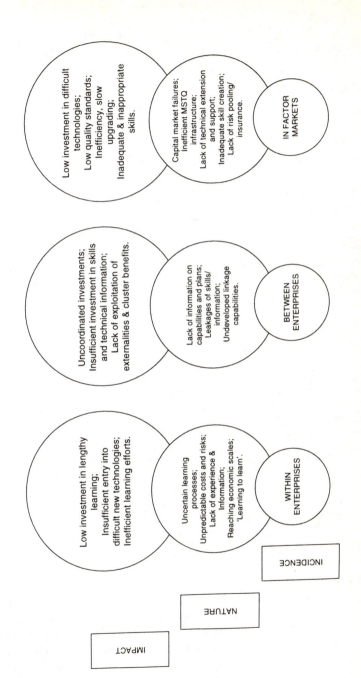

IMPACT

NATURE

INCIDENCE

WITHIN ENTERPRISES

Low investment in lengthy learning;
Insufficient entry into difficult new technologies;
Inefficient learning efforts.

Uncertain learning processes;
Unpredictable costs and risks;
Lack of experience & information;
Reaching economic scales;
'Learning to learn'.

BETWEEN ENTERPRISES

Uncoordinated investments;
Insufficient investment in skills and technical information;
Lack of exploitation of externalities & cluster benefits.

Lack of information on capabilities and plans;
Leakages of skills/ information;
Undeveloped linkage capabilities.

IN FACTOR MARKETS

Low investment in difficult technologies;
Low quality standards;
Inefficiency, slow upgrading;
Inadequate & inappropriate skills.

Capital market failures;
Inefficient MSTQ infrastructure;
Lack of technical extension and support;
Inadequate skill creation;
Lack of risk pooling/ insurance.

Figure 1.2   Constraints on enterprise development

- Distortions introduced by interventions must be offset. In particular, protection must be accompanied by competitive pressures to enter world markets. This is what traditional I-S strategies failed to provide.

- Since intervention resources are limited, only a few activities should be supported at any time. Intervening in a large number of unrelated activities risks waste and failure.

- Since learning is a cumulative and incremental process, interventions must aim to support activities that have a base in existing skills and knowledge in a country. New technological 'leaps' must be modest, based on realistic assessment of what is feasible within reasonable periods of time.

- The line between market-friendly and selective interventions is almost impossible to draw. Each market may be subjected to a combination of functional and selective policies. Figure 1.3 shows some of the interventions in the main markets within which capability development occurs (drawn from the actual policies of the NIEs).

THE EAST ASIAN MODEL

It is now widely accepted that there *was* no 'East Asian model' of industrialisation. There is a different model for each NIE, within a common context of export orientation, good human capital and strong regional spillovers. Each NIE had different industrial objectives and used different interventions (though some, like support for exporters and for small enterprises, were very similar). As a result, each had a different pattern of industrial and export growth, reliance on FDI, technological capability and enterprise structure. However, for none, even the least interventionist, was simply 'getting prices right' a sufficient explanation of industrial success. The different objectives of the NIEs are shown in a simplified form in Table 1.1. There was an enormous range, from the *laissez faire* to selective targeting and control.

Hong Kong

Hong Kong was at one end of the range, combining free trade, no selective targeting and an open-door policy to FDI. An object lesson in the virtues of free trade to other developing countries? Not necessarily: Hong Kong had unique initial conditions – its long *entrepôt*

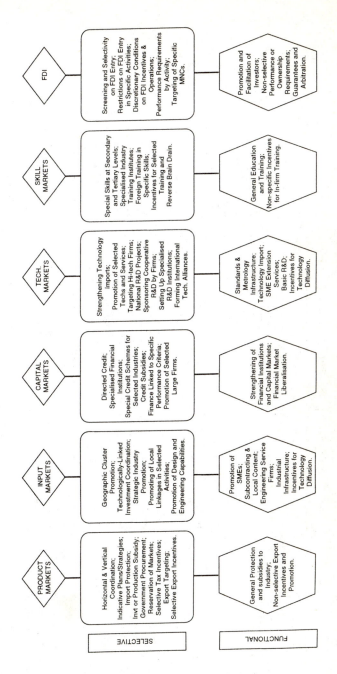

Figure 1.3   Selective and functional interventions in various markets

Table 1.1 Industrial policy objectives of NIEs

	Deepening industrial structure	Raising local content	FDI strategy	Raising technological effort	Promotion of large local enterprises
Hong Kong	None	None	Passive Open Door	None except tech. support for SMEs	None
Singapore	Very strong push into specialised high skill/tech industry, without protection	None, but subcontracting promotion now started for SMEs	Aggressive targeting & screening of MNCs, direction into high value-added activities	None for local firms, but MNCs targeted to increase R&D	None, but some public sector enterprises enter targeted areas
Taiwan	Strong push into capital-, skill- and technology-intensive industry	Strong pressures for raising local content and subcontracting	Screening FDI, entry discouraged where local firms strong. Local technology diffusion pushed	Intense tech. support for local R&D & upgrading especially by SMEs. Govt orchestrated high-tech development	Sporadic: to enter heavy industry, mainly by public sector
Korea	Strong push into capital-, skill- and technology-intensive industry, especially heavy intermediates and capital goods	Stringent local-content rules, creating support industries, protection of local suppliers, sub-contracting promotion	FDI kept out unless necessary for technology access or exports, joint ventures and licensing encouraged	Ambitious plans for local R&D in advanced ind., heavy investment in technology infrastructure. Targeting strategic technologies	Sustained drive to create giant private conglomerates to internalise markets, lead heavy industry, create export brands

Table 1.2   Manufactured exports by NIEs ($m. and annual growth rate)

		Hong Kong		Korea		Singapore		Taiwan	
		1987	1992	1987	1992	1987	1992	1987	1992
Total	Value	44597	28738.5	43398.2	71046.4	20586.2	49441.1	47276.6	75643.4
	Growth	−8.4%			10.4%		19.2%		9.9%
Textiles &	Value	11537.1	8591.3	6957.9	8097.8	497.5	824	5444.4	6035.2
Clothing	Growth	−5.7%			3.1%		10.6%		2.1%
Machinery & Transport Equipment	Value	10664.5	7260.2	15566.8	30557.6	12294.6	32960.7	15250.5	32534.8
	Growth	−7.4%			14.4%		21.8%		16.4%

Source: World Development Report, 1989 and 1994.

tradition (with global trading links), an established infrastructure of trade and finance, the presence of large British companies (the 'Hongs') with immense spillovers in skills and information, and an influx of entrepreneurs and trained textile and metalworking engineers and technicians (with considerable embodied learning) from mainland China. This unique background allowed it to launch into export-oriented light manufacturing under free trade, and its experience has been unique in the developing world. Simply 'getting prices right' has not created other Hong Kongs, even in other free-trade havens with good human capital and favourable location.

Moreover, the lack of selective promotion had important effects on the manufacturing structure. Hong Kong *started and stayed with* light labour-intensive manufacturing industry,[13] where learning costs were relatively low and predictable. There was some 'natural' progression as product quality improved and new consumer products were added. But there was little industrial or technological deepening over time, in contrast to the other NIEs that pursued selective deepening strategies.

As a result, Hong Kong underwent massive deindustrialisation as wages and land costs rose (during 1986–92 it lost about 35 per cent of its manufacturing employment, and the process is continuing[14]). The colony relocated its manufacturing to other countries, mainly China, and its own export growth went into decline after the mid-1980s (Table 1.2). The economy is continuing to grow and prosper, but the lessons of the Hong Kong 'miracle' for *industrial development* are ambiguous. The lack of industrial deepening and deindustrialisation, a direct result of the absence of industrial policy, would be very undesirable in other developing economies.

Singapore

In contrast, Singapore illustrates the consequences of an intervention-ist policy combined with free trade. Singapore has half the population of Hong Kong and higher wages, but has not suffered a similar hol-lowing out. Its industrial structure is far deeper (in the complexity of production and exports), and it enjoys high sustained industrial growth. It relies heavily on MNCs; but, unlike Hong Kong, the government has targeted activities for promotion and aggressively sought and used FDI as the tool to achieve its objectives. Singapore started with a base of capabilities in *entrepôt* trading, ship servicing and petroleum refin-ing. After a brief period of import substitution, it moved into export-oriented industrialisation, based overwhelmingly on investment by MNCs. There was little influx of technical and entrepreneurial know-how from China, and a weak tradition of local entrepreneurship. After a decade or so of light industrial activity (garment and semiconductor assem-bly), the government acted firmly to upgrade the industrial structure. It guided MNCs to higher value-added activities, narrowly specialised and integrated into the world-wide structure of their operations. It in-tervened to create the specific skills needed,[15] and set up public enter-prises to undertake activities considered in the country's future interest, where foreign investment was unfeasible or undesirable.

Such specialisation, along with the heavy reliance on foreign invest-ments, greatly reduced the need for indigenous technological effort. While the government mounted strong efforts to induce MNCs to es-tablish R&D facilities, the technological depth of the affiliates is still comparatively low. This technological strategy is feasible only for rela-tively small and specialised economies, and may not be relevant to most developing countries with a large local industrial structure and a more diverse range of activities.

Korea and Taiwan

Korea and Taiwan adopted far more interventionist strategies on trade and domestic resource allocation. They had a clear preference for pro-moting indigenous enterprises and for deepening local technological capabilities, and assigned FDI a secondary role to technology import in other forms. Their export drive was led by local firms, and a series of interventions allowed local firms to develop impressive technologi-cal capabilities. The domestic market was not exposed to free trade; a range of quantitative and tariff measures were used over time to give infant industries 'space' to develop their capabilities. The deleterious effects of protection were offset by strong incentives (in the case of Korea, almost irresistible pressures) to export and face full international competition.

Korea

Korea went much further in developing advanced and heavy industry than Taiwan.[16] To achieve its compressed entry into heavy industry, its interventions had to be more detailed and pervasive. Korea relied primarily on capital-goods imports, technology licensing and other technology-transfer agreements to acquire technology. It used reverse engineering, adaptation and own-product development to build upon these arm's-length technology imports and develop its own capabilities. Its R&D expenditures are now the highest in the developing world, and ahead of all but a handful of leading OECD countries (see Chapter 3 on technology development).

One of the pillars of Korean technological strategy, and one that marks it off from the other NIEs (but mirrors Japan), was the deliberate creation of large private conglomerates, the *chaebol*. The *chaebol* were hand-picked from successful exporters and were given a range of subsidies and privileges, including the restriction of MNC entry, in return for furthering the strategy of setting up capital- and technology-intensive activities geared to export markets. The rationale for fostering size was obvious: in view of deficient markets for capital, skills, technology and even infrastructure, large and diversified firms could internalise many of their functions. They could undertake the cost and risk of absorbing very complex technologies (without a heavy reliance on FDI), further develop it by their own R&D, set up world-scale facilities and create their own brand names and distribution networks. This was a costly and high-risk strategy; the risks were contained by the strict discipline imposed by the government in terms of export performance, vigorous domestic competition, and deliberate interventions to rationalise the industrial structure. The government also undertook various measures to encourage the diffusion of technology, putting pressures on the *chaebol* to establish vendor networks.[17] Apart from the direct interventions to support local enterprises, the government provided selective and functional support by building a massive technology infrastructure and creating general and technical skills. Korea today has the highest rate of university enrolment in the developing world, and produces more engineers each year than the whole of India.

Taiwan

Taiwan's industrial policy encompassed import protection, directed credit, selectivity on FDI, support for indigenous skill and technology development and strong export promotion.[18] While this resembles Korean strategy in many ways, there are important differences. Taiwan did not promote giant private conglomerates, nor did it attempt the intense drive into heavy industry that Korea did. Taiwanese industry is largely composed of SMEs, and, given the disadvantages to technological activity inherent in small size, these were supported by a variety of inducements and institutional measures in upgrading their technologies. Taiwan has the developing world's most advanced system of technology support for SMEs.

In the early years of industrialisation, the Taiwanese government attracted FDI into activities in which domestic industry was weak, and used a variety of means to ensure that MNCs transferred their technology to local suppliers. As with Korea, FDI was directed to areas where local firms lacked technological capabilities. The government also played a very active role in helping SMEs to locate, purchase, diffuse and adapt new foreign technologies. Where necessary, the government itself entered into joint ventures, for instance to get into technologically very difficult areas such as semiconductors and aerospace.

This sketch of the policies of the NIEs leads to the following conclusions:

- Selective as well as functional interventions played a vital role in the pattern of industrial and technological development in the NIEs. The extent of industrial and technological deepening achieved was strongly related to selective interventions to promote such deepening.

- Governments showed an ability to devise and implement interventions effectively, partly because export orientation imposed a strict discipline on both industry and governments and partly because of the high levels of training, adequate remuneration and political insulation of bureaucrats.

- The nature and impact of interventions differed according to differing government objectives and political economies.

- FDI was treated very differently by each of the four countries and so played very different roles in their technological development. Those that wanted to promote *indigenous* technological deepening had to intervene to restrict foreign entry and to guide their activities and maximise the spillovers. Those that chose to rely on MNCs and upgrade within their global production structure had to intervene to target investors, guide their allocation and induce them to set up more complex functions than they would otherwise have done.

- The options and compulsions applicable to the larger economies, with greater scope for internal specialisation and local content as well as better established indigenous enterprises, were different from those open to small states with weak indigenous entrepreneurship and a tiny internal market. Given the need to spread technological development more widely, the former had to take more direct steps to assist local firms.

LESSONS FOR POLICY

While it may be accepted in principle that interventions can be helpful to remedy market failures, the World Bank's 'market-friendly' approach reflects a strong ideological strain in economics that governments are intrinsically unable to act (selectively) in the national interest.[19] There may be several reasons: governments cannot have enough information to select better than the market; they do not have the skills to design and implement detailed interventions; they are inflexible and unable to change course when mistakes become apparent; they tend to represent sectional rather than national interests; and they are venal or corruptible.

These reasons have some validity. There are clearly circumstances in which particular governments cannot be entrusted with selectivity. However, these are not absolute givens that rule out selectivity altogether. Let us briefly consider the arguments:

- *Lack of information*: Most developing-country governments lack the information to make selective decisions. However, difficulties in 'picking winners' can be exaggerated. Industrial latecomers have much more information (on market and factor conditions, technological requirements, skill and organisational needs) than countries at the frontiers of innovation, where the risk of selectivity is much greater. It is easier for the former to follow countries further up the industrialisation scale: the way that Korea 'followed' Japan is a case in point. Moreover, industrial policy does not involve picking winners so much as *creating* them. There are a number of viable options facing late industrialisers, any of which could be made to work if the right skills, technologies, institutions and incentives are mustered. What is necessary is to be 'right' in a broad range, and to mount a systematic and coherent strategy. Finally, where information is lacking, there is certainly a need for governments to collect it, from other countries, from domestic sources and by close interaction with the industrial sector. One of the most important lessons of the Asian NIEs, well analysed by the *Miracle* study, is that interventions were not conducted by bureaucrats acting on grand plans based on abstract planning models or grandiose schemes of national aggrandisement, but in close consultation with the private sector. This provided information on trends and conditions that the government could not have accessed otherwise. Note also that they exercised different levels of selectivity – Korea was much more detailed and pervasive than Taiwan, and called for more detailed information. Lower levels of selectivity are less information-intensive, and also involve lower risks.

- *Skills*: Many administrations do not at present have the economic or technical skills to design and mount selective interventions, as in the least developed countries of sub-Saharan Africa. Perhaps more important, they are often given multiple, unclear or conflicting objectives which make it difficult to design and monitor policies. These can be remedied, albeit slowly, by increasing the education and training base of the economy and by having clearer economic objectives (the point about levels of selectivity also applies here). The Asian NIEs had the clear objective of increasing exports and gaining international competitiveness. This enabled them to design policies and to deploy their skills much more effectively.

- *Inflexibility*: Many interventions turn out to be costly not so much because they are poorly designed (private business makes huge mistakes all the time) but because changing course is difficult and there is no official accountability for the outcome. Clearly all interventions have to be designed flexibly and monitored constantly so that mistakes can be rectified as they become apparent. There are precedents in the private corporate sector on how this can be done, but perhaps the most effective check is to impose performance requirements (e.g. export growth) and to make officials more directly accountable. Export orientation was itself the best guarantee of flexibility in policy making in the NIEs (Moreira, 1994). If the intervention is kept at fairly general levels, the 'tie in' to particular choices is also correspondingly lower and the task of changing direction easier.

- *Sectional interest*: 'Sectionalisation' of decision making is a danger in most governments. This affects functional as well as selective interventions, of course, but the dangers may be greater for the latter. It can only be offset by clear leadership, the setting up of appropriate institutions and internal checks on the allocation of favours – that this can be done is amply illustrated by the Asian experience.

- *Corruption*: There are several levels to this problem: the higher the level the more difficult it is to solve. At lower levels of government, changes in monitoring, employment conditions, salaries and incentives may help reduce rampant corruption. At the top levels, no one is able to impose sanctions on wrong-doers. Again, such venality can distort the most liberal regime, not just interventionist ones. The solutions, if any, lie in larger political and social processes that are beyond the purview of this analysis, but certainly a corrupt government should not be entrusted to undertake detailed industrial policy.

It does not appear, therefore, that the objections amount to a universal and permanent case against selectivity. The question is more of degree than of kind. There are *some* levels of selective intervention that most governments can undertake, and there are some governments that cannot for the time being be entrusted with any – but these are governments that are unlikely to carry through even the market-friendly interventions that all development requires.

Some new political economists hold that there are no circumstances in which any government can be trusted to act impartially in the national interest and do it effectively. This is biased and ideological – it is not supported by the evidence of East Asia, and it is not clear that it would hold up to historical evidence in the West. Governments are fallible (just as markets are), but they can be improved. Government structures can be reformed, skills created, impartiality increased. It is only corruption and venality that perhaps is difficult to remove by an act of will, but then corrupt governments exist in liberal economies and free markets do not remove rent seeking.

There are degrees of industrial policy, with different levels and detail of selectivity in intervention. The need for industrial policy can also change with the development of markets; as economies develop and markets grow more competent and sophisticated, the need for intervention diminishes. What is important to remember is that *not intervening has its own costs*. Market failures can stunt industrialisation if all governments do is 'get prices right', and wait for markets to do the rest. Even market-friendly interventions combined with liberal policies can narrow and constrict industrial development. The lesson of the larger NIEs is precisely that these constraints of the market can be relaxed, and the industrialisation process greatly compressed and dynamised, by appropriate interventions. Countries need not be satisfied with the market-given pace and content of industrial development, but can *use* the market to enlarge their opportunities.

Is the East Asian case replicable? Not perhaps in all its ramifications: no other country can 'be a Korea' in the details of strategy. But then Korea was not Japan, and Taiwan was not Korea. There were sufficient similarities in their approach to identifying and remedying market failures that offer generic lessons for the rest of the developing world. These lessons are not only economic; they also concern the design, administration, financing and staffing of interventions. This is where the *Miracle* study is particularly good, though after discussing the ways in which interventions were designed, 'contests' set up and neutrality promoted, it concludes that these were unique to the East Asians. This seems mistaken, if not patronising and offensive. Different economic, institutional and political conditions certainly dictate different strategies, but they do not rule out strategies altogether.

REFERENCES

Amsden, A. H., 'Why Isn't the Whole World Experimenting with the East Asian Model to Develop? Review of *The East Asian Miracle*', *World Development*, vol. 22, no. 4 (April 1994), pp. 627–34.

Amsden, A. H., *Asia's Next Giant: South Korea and Late Industrialization* (New York: Oxford University Press, 1989).

Arrow, K. J., 'Economic Welfare and the Allocation on Resources for Innovation', in R. Nelson (ed.), *The Rate and Direction of Innovative Activity* (Princeton: Princeton University Press, 1962), pp. 609–26.

Bell, M. and Pavitt, K., 'Technological Accumulation and Industrial Growth: Contrasts between developed and developing countries', *Industrial and Corporate Change*, vol. 2, no. 2 (1993), pp. 157–210.

Brautigam, D., 'The State as Agent: Industrial development in Taiwan, 1952–1972', in H. Stein (ed.), *Asian Industrialization and Africa* (London: Macmillan, 1994), pp. 145–82.

Chang, Ha-Joon, *The Political Economy of Industrial Policy* (London: Macmillan, 1994).

Fishlow, A., Gwin, C., Haggard, S., Rodrik, D. and Wade, R., *Miracle or Design? Lessons from the East Asian Experience* (Washington, DC: Overseas Development Council, 1994).

Jacobsson, S., 'The Length of the Learning Period: Evidence from the Korean engineering industry', *World Development*, vol. 21, no. 3 (March 1993), pp. 407–20.

Katz, J. M. (ed.), *Technology Generation in Latin American Manufacturing Industries* (London: Macmillan, 1987).

Kim, K. S., 'The Korean Miracle (1962–80) Revisited: Myths and realities in strategies and development', in H. Stein (ed.), *Asian Industrialization and Africa* (London: Macmillan, 1994), pp. 87–144.

Kwon, J. 'The East Asian Challenge to Neoclassical Orthodoxy', *World Development*, vol. 22, no. 4 (April 1994), pp. 635–44.

Lall, S., 'Governments, Globalization and International Business: The developing East Asian economies', in J. H. Dunning (ed.), *Governments, Globalization and International Business* (forthcoming).

Lall, S., '"*The East Asian Miracle*" Study: Does the bell toll for industrial strategy?', *World Development*, vol. 22, no. 4 (April 1994.a), pp. 645–54.

Lall. S., 'Industrial Policy: The role of government in promoting industrial and technological development', *UNCTAD Review 1994* (1994.b), pp. 65–89.

Lall, S., 'Understanding Technology Development', *Development and Change*, vol. 24, no. 4 (October 1993), pp. 719–53.

Lall, S., 'Technological Capabilities and Industrialization', *World Development*, vol. 20, no. 2 (February 1992), pp. 165–86.

Lall, S., *Building Industrial Competitiveness in Developing Countries* (Paris: OECD, 1990).

Lall, S. and Najmabadi, F., *Bank Lending for Industrial Technology Development* (Washington, DC: World Bank, Operations Evaluation Department, 1995).

Lim, L., 'Foreign Investment, the State and Industrial Policy in Singapore', in H. Stein (ed.), *Asian Industrialization and Africa* (London: Macmillan, 1994), pp. 205–38.

Lucas, R. E., 'On the Mechanics of Economic Development', *Journal of Monetary Economics*, vol. 22, no. 1 (March 1988), pp. 3–42.

Mill, J. S., *Principles of Political Economy* (London: Longmans, Green and Company, 1940; first published in 1848).

Moreira, M. M., *Industrialization, Trade and Market Failures: The Role of Government Intervention in Brazil and the Republic of Korea* (London: Macmillan, 1994).

Nelson, R. R., 'Research on Productivity Growth and Productivity Differences: Dead ends or new departures?,' *Journal of Economic Literature*, vol. 19, no. 3 (June 1981), pp. 1029–64.

Nelson, R. R. and Winter, S. J., *An Evolutionary Theory of Economic Change* (Cambridge, Mass.: Harvard University Press, 1982).

OED, *World Bank Support for Industrialization in Korea, India and Indonesia* (Washington, DC: World Bank, Operations Evaluation Department, 1992).

Pack, H. and Westphal, L. E., 'Industrial Strategy and Technological Change: Theory versus reality', *Journal of Development Economics*, vol. 22, no. 1 (September 1986), pp. 87–128.

Rosenberg, N., *Perspectives on Technology* (Cambridge: Cambridge University Press, 1986).

Selvaratnam, V., 'Innovations in Higher Education: Singapore at the competitive edge' (Washington, DC: World Bank, 1994), Technical Paper no. 222.

Shapiro, H. and Taylor, L., 'The State and Industrial Strategy', *World Development*, vol. 18, no. 6 (June 1990), pp. 861–78.

Singh, A., 'Openness and the Market Friendly Approach to Development: Learning the right lessons from development experience', *World Development*, vol. 22, no. 12 (December 1994), pp. 1811–24.

Stiglitz, J. E., 'Learning to Learn, Localized Learning and Technological Progress', in P. Dasgupta and P. Stoneman (eds), *Economic Policy and Technological Development* (Cambridge: Cambridge University Press, 1987), pp. 125–55.

Stiglitz, J. E., 'Markets, Market Failures and Development', *American Economic Review Papers and Proceedings*, vol. 79, no. 2 (1989), pp. 197–202.

Stiglitz, J. E. *et al.*, *The Economic Role of the State* (Oxford: Basil Blackwell, 1989).

Streeten, P. P., 'Markets and States: Against minimalism', *World Development*, vol. 21, no. 8 (August 1993), pp. 1281–98.

Wade, R., *Governing the Market: Economic Theory and the Role of Government in East Asian Industrialization* (Princeton: Princeton University Press, 1990).

Westphal, L. E., 'Industrial Policy in an Export-Propelled Economy: Lessons from South Korea's Experience', *Journal of Economic Perspectives*, vol. 4, no. 3 (March 1990), pp. 41–59.

World Bank, *The East Asian Miracle: Economic Growth and Public Policy* (New York: Oxford University Press, 1993).

Young, A., 'Learning by Doing and the Dynamic Effects of International Trade', *Quarterly Journal of Economics*, vol. 56, no. 2 (May 1991), pp. 369–405.

2 Understanding Technology Development

INTRODUCTION

In a general sense, most developing countries are inept at using industrial technologies.[1] This ineptness may take one or more of several forms. Their enterprises may import and deploy technologies that are not appropriate to their countries' endowments of labour or skills, or to their scales of production. They may not fully assimilate the technologies they have imported, so may not use them at 'best-practice' levels of technical efficiency. Individual enterprises may, moreover, differ widely in their relative efficiencies. This variation is also found in developed economies, but tends to be far more marked in developing ones, indicating that diffusion of knowledge is more limited and market competition more imperfect. Finally, they may not be able to upgrade the technologies they have mastered, or diversify into new technologies as conditions change. Thus, they may stay at the low value-added end of the industrial spectrum, falling behind world technological frontiers as others forge ahead.

There are, however, significant differences between developing countries in this respect. A few advanced NIEs are approaching, and in a few cases surpassing, established best-practice norms for given technologies. They are proving able to keep up with world technological frontiers in many complex technologies. In contrast, there are many countries that are unable to deploy even simple technologies efficiently, and show few signs of technological dynamism. In between lie the majority of developing countries, with a mixture of poor and good performances. Yet all these countries have access to the same world stock of knowledge and equipment, and may even have bought their technologies from the same sources.

Part of this difference in performance is traceable to different policy environments that allow firms to respond properly to market incentives in one country and hold the response back in another. Much of the rest is, however, explained by differences in the ability of productive enterprises to handle industrial technologies and cope with technical change: by their *technological capabilities*. Technological capabilities (TCs) in industry are the skills – technical, managerial and institutional – that allow productive enterprises to utilise equipment and technical information efficiently.[2] Such capabilities are firm-specific, a form of institutional knowledge that is made of the combined skills of its members accumulated over time. The development of such capabilities may be defined as *industrial technology development* (ITD).

ITD in developing countries should not be thought of as the development of the ability to undertake frontier innovation, though innovative capabilities are one form of TC. It comprises a much broader range of effort, reviewed below, that every enterprise must itself undertake in order to absorb and build upon the knowledge that has to be utilised in production. The successful transfer of a new technology to a developing country thus has to include a major element of capability building: simply providing equipment and operating instructions, patents, designs or blueprints does not ensure that the technology will be properly used. These 'embodied' elements of a technology must be accompanied by a number of 'tacit' elements, which have to be taught and learned.[3]

The technology market is thus different from a product market, and it is important to understand the differences. Unlike the sale of a good, where the transaction is completed when the physical delivery takes place, the transfer of skills and information is a prolonged process. It involves local learning to complete the transaction, and the recipient of the technology has to invest in absorbing the new knowledge. Furthermore, industrial enterprises do not usually develop capabilities in isolation. They draw upon factor markets, institutions and other enterprises for skills, assistance and information. The markets for finance for technology development, for the creation of new skills and for the generation and diffusion of technical information tend to suffer from imperfections, even in developed countries. In developing countries the fragmentation, gaps and externalities inherent in these markets are generally larger.[4] Left to their own devices, individual enterprises may find capability development very difficult, slow and expensive to undertake, and so may end up with poor abilities to operate efficiently or upgrade their technologies.

The remedying of failures in markets for finance, skills and information creates a strong case for government intervention, in developed as well as developing countries. The technology development in question refers to the growth and strengthening of TCs, not to the support of high-tech R&D (which is the form it normally takes in advanced industrial countries). Even such a modest aim has many ramifications. Its proper design and implementation requires a range of administrative and political capabilities. Where governments lack these capabilities, policies to promote the development of industrial capabilities have to be correspondingly limited in scope.

This chapter sketches out the case for industrial technology-development policies, and the shape of appropriate policies, in developing countries. It starts by describing the characteristics of capability development, followed by the determinants of TC. The policy needs for ITD are given by the nature of failures in the relevant markets, but a realistic policy analysis must take account of the risk, strong in many developing countries, of government failure, or at least of limited government capabilities. The final section draws the main conclusions.

CHARACTERISTICS OF TECHNOLOGICAL CAPABILITIES

To reiterate, TCs in manufacturing are the skills, technical knowledge and organisational coherence required to make industrial technologies function in an enterprise.[5] They are not the technology that is 'embodied' in physical equipment or in manuals, blueprints and patents that are purchased by the enterprise, though these are the tools with which capabilities are put to work. Nor are they simply the educational qualifications possessed by the employees, though the receptive base for the acquisition of capabilities depends to a large extent on the education and training of the people involved. They are not even the skills and learning undergone by individuals in the enterprise, though these are the building blocks of capability building at the micro level. They are the way in which an institution like a firm combines all the above *to function as an organisation*, with constant interaction among its members, effective flows of information and decisions, and a synergy that is greater than the sum of individual skills and knowledge.[6]

To some extent any enterprise that tries to use a new technology acquires TC as an automatic result of the production process. This passive 'learning by doing' goes some way to developing the necessary capabilities. In simple industries, say the assembly of imported kits or garment manufacture for the domestic market, this may be all that is needed. There is little need to invest in creating new skills in the workforce or management, or in investigating ways of lowering costs by substituting materials, saving energy, searching for new designs and so on. The skills that are needed are easily learned on the job, and factor markets are generally uncomplicated (there are few interlinkages with suppliers that involve technical problems and complex exchanges of information). Product designs are given by foreign suppliers of kits for assembly, or are easily adapted to local tastes in garments.

Such passive learning is, however, insufficient as the technology becomes more complicated or market demands more rigorous. Even in garment manufacture, for instance, a lot of effort is needed to raise quality and productivity, improve layout, introduce new supervision practices and so on, before an efficient producer for the local market becomes a competitive exporter (even when subcontracting for a buyer who provides all the designs and specifications). For more complex industries, to reach even static 'best-practice' levels as established in advanced countries involves an enterprise in a longer and more demanding process.

Where the technology is new to the firm, it has to search for and hire new skills from the market. If these are available and adequate, there are generally a number of tasks that have to be undertaken to make the technology operational: taking 'bugs' out of the plant, adapting the technology to local scales, inputs and demand conditions, learning the elements of product design or of process and layout optimisation, building the basic procurement networks, and so on. Where the technology is new to the country, these skills may not be available locally. It may then have to import the skills: but this is expensive and local skills have to be developed quickly, not just formally but in a way that makes them functional in a plant (with on-the-job experience).

The learning process for complex technologies can thus be long and uncertain. In a new environment, with new plant or process, different physical conditions, and fresh workers and managers, no technical process can be entirely predictable. Thus, even the achievement of static efficiency (the solution, in other words, of the problems mentioned above) usually involves search, experimentation and interactions within firms as well as between them. When dynamic factors are introduced, the process of becoming and staying efficient is even more difficult.[7] The real world is in a state of constant flux. Market conditions and tastes are changing, technologies improving, new competitors appearing, and relative costs of inputs, labour of various kinds and infrastructure are shifting. Any firm that is to stay competitive has to invest in constant skill development and technological effort. The nature of the effort will differ by the nature of competition and efficiency of factor and product markets (see below), but success will generally depend on constant investment in capability acquisition.

It is useful conceptually to look at the development of firm-level competitiveness as *investment in 'embodied' technology* (plant, equipment, licences, blueprints, other external inputs) accompanied by *investments in skills, information, organisational improvements and linkages* with other firms and institutions. If investments in embodied technology outpace investments in TC, the results are inefficiency, stagnation and waste. The reverse is less likely, since TCs are generally acquired on the job, but it often happens that distorted incentive regimes hold back the full exploitation of TCs acquired over long periods of learning in the past (as in many inward-oriented economies with a strong base of technical skills and institutions, such as that of India[8]). In general, however, in most developing countries there are significant gains to be made by improving capabilities without adding greatly to the stock of embodied technology.

Regarding capability building as an investment helps highlight its most important features. In brief, these are:

- The investment has to be *conscious and purposive*. Only a very limited range of capabilities can be developed by a process of costless, automatic learning as an accompaniment to the production process. Many firms can go on producing for years without acquiring the skills and knowledge to operate efficiently.

- Investments in TC, like other investments, are *highly sensitive to the incentive environment, the cost of the investment, and the availability of the investible resources* (apart from the finance, these are access to skills, technical information and support, and interactions with related firms). Thus, certain economies experience greater TC investments than others because of their more stable and conducive macroeconomic policies, more stimulating market environment, greater stocks of the relevant skills, and better technology institutions (see below on the determinants of TC).

- There is *no predictable learning curve* down which all firms travel. In any given environment, some firms will invest more in TC than others, depending on their size, information, strategy, risk aversion and access to resources. In every economy, developed or developing, there are wide and persistent differences in firm-level efficiency, depending on differing TC investments. It was noted above that the narrowing of these differentials, by bringing the poorer firms nearer the frontier, can be a major source of productivity increase in developing countries. Random entrepreneurial differences apart, the extent to which individual firms invest in TC depends on the efficient functioning of factor and information markets, and the realisation by the entrepreneur that TC investments are needed and can be profitable. In most developing countries, *the learning process itself has to be learnt.*

- The growth of TC can take place at almost *any point in the production process*: shop-floor, engineering, design, procurement or formal R&D.[9] It would be mistaken to associate technological development only with formal R&D. The relative importance of investing in different capabilities depends on the nature of the technology and the level of TC development already reached by the firm. Many of what are regarded as routine functions in developed countries (quality control, maintenance, process optimisation, inventory control, and the like) are sources of potential productivity gains in developing ones, and should attract relatively more investment in the latter. However, R&D becomes important as a form of TC investment, even in developing countries, for the absorption of very advanced technologies. In the NIEs, where some nationally-owned enterprises are approaching world technological frontiers, autonomous R&D capacity becomes essential. This is so even in medium-technology industries because firms find that they cannot continue to depend on purchases of new technology from advanced countries.

- The complexity, cost and risks of TC investments rise with the *sophistication of the technology and the level of technological development* reached. The cost of building up a viable level of TC in, say, automobile manufacturing is far greater than in the making of simple machine tools, which in turn is more than in garments. The process of industrial development and the dynamic evolution of comparative advantage in manufacturing involves the deepening of TCs in given technologies and diversification into more complex technologies. The optimal allocation of TC investments cannot be predicted for developing countries in general since the initial conditions differ greatly, and since government policies can have a major impact on the pace and direction of capability building.

- Manufacturing enterprises do not develop capabilities in isolation. They operate, in many industries, in a *dense network of formal and informal relationships* with suppliers, customers, competitors, consultants, and technology, research and educational institutions. These networks take the form of complex contractual and non-contractual relations. Over time, they tend to lead to the establishment of institutions and 'rules of the game' by industries or the government to foster and regulate these linkages and externalities. These help individual firms to deal with each other, to gain access to expensive ('lumpy') information and facilities, and to create information, skills and standards that all firms need but no individual firm will generate (what are termed 'public goods'). The need for such linkages differs by industry and enterprise. Some simple activities that rely mainly on primary inputs and use low-skill techniques may not need many linkages to be efficient. Others need constant flows of information, skills and other inputs from outside to maintain their efficiency and to make rational decisions to invest in physical and human assets. As the role of complex activities grows with economic development, so does the need to take account of linkages by proper coordination of infrastructural and institutional measures.

• Finally, the development of TCs in enterprises, and so in industries and economies, follows an *evolutionary but individual path*.[10] Investments in knowledge and skill creation are cumulative, building upon past investments. Easier capabilities and activities are mastered first, then more difficult but related ones, and so on. However, within this broad pattern there are many deviations. At the enterprise level, individual firms operate with the limited amount of technological knowledge that they have gathered related to the technologies they use and the modifications they have made to them. They are, in other words, familiar with a point on the production function rather than the complete curve. They make 'idiosyncratic' changes and progress in individual ways rather than moving uniformly to take advantage of innovation possibilities given externally. Similarly, developing economies follow an evolutionary path based on their initial positions, but deviate considerably over time as their firms grow, policies vary, markets improve and external factors have random impacts.

HOW THIS DIFFERS FROM TEXTBOOK MODELS

The process of firm development and efficiency described here differs considerably from the textbook economic model of this process. The theoretical model is presented as a simplified version of reality, but its assumptions are believed to capture the essential features of what happens in reality. These assumptions are the bedrock of much of the current understanding of technology and industrial development in developing countries.[11] They also provide the rationale for important elements of policy making and policy advice in this area. It is useful, therefore, to quickly run through them to see some of the contrasts with the TC approach.

In the world of simple theory, firms are assumed to operate with full knowledge of all possible technologies (i.e. on a well specified production function that is known to all countries). Given the right market-determined prices for inputs and outputs, they pick the one that is appropriate to their national factor endowments. All firms in an industry facing the same prices choose the same technologies; otherwise they are allocatively inefficient. Technology transfer is like the sale of a product. There are no 'tacit' elements in the transfer, no learning costs and no need to make adaptations. Thus, all firms can immediately use technologies with the same degree of efficiency (and all at best-practice levels). Technical inefficiency is, *ex hypothesi*, due to managerial slack or incompetence, since there are no learning costs.

There is taken to be little technological activity in developing countries: most significant technological activity is supposed to be related to 'innovation', which is something that only developed countries undertake. Where extra costs of introducing new technologies in developing countries are admitted, a simplified view is taken of the learning process. The learning curve is believed to be fairly short and predictable. It is confined to 'running in' a new plant until it reaches rated capacity, and until the benefits of shop-floor learning by repetitive activity are realised. It is generally assumed that such costs are relatively trivial and similar across industries. The learning process is, moreover, relatively passive rather than one that involves investment, risk and, in some activities, long maturation periods. Firms know what to do to reach best-practice levels; there is no 'learning to learn' problem. There is consequently no need to devise measures to stimulate the capability to absorb new technologies, or to distinguish between industries in the financing of learning and start-up costs.

Firms are assumed to acquire and use technologies as individual units, essentially in isolation. There are no linkages between them, and no externalities resulting from individual efforts to generate skills and information. The development of specialisation among firms and industries relies solely on information exchanged in anonymous market transactions. There is no need to create or foster the information networks and institutions that have evolved in advanced industrial economies. Since there are no technological externalities, there is also no need to coordinate investment decisions (in physical assets or TC building) across activities that may have intense linkages. Nor are some sets of activities more significant for industrial development than others because they have more beneficial externalities: no sector has the claim to be more 'strategic' than another.

Where learning costs (of the minor sort noted above) are admitted, it is assumed that firms have the foresight to finance these predictable costs in capital markets. If capital markets are not fully efficient, infant-industry protection may be granted as a second-best measure. However, any protection must be at moderate levels and must be uniform across activities to minimise the risk of misallocation of investments. Since no technologies are more difficult, or involve more market failures due to externalities, than others, there is no need for policy makers to be selective in their promotion of particular industries or technologies. There is also no need to devise different policies by the level of development of the countries in question. A uniform approach across countries, as well as within them, is the least distortionary.

In this essentially static framework, comparative advantage is determined by the accumulation of 'factor endowments' broadly defined, rather than by the deliberate TC-raising efforts of industrial enterprises. As endowments grow, firms automatically move to the right factor combinations, costlessly absorbing the technologies concerned. It is generally admitted, however, that in developing countries some factor markets may not operate efficiently, in particular in the creation of skills and the diffusion of information. In this case policy interventions are needed to support factor markets. But this support has to be non-selective. The allocation of resources should not be influenced to encourage particular activities that governments consider desirable (because they raise the TC of the industrial sector, have greater linkages or face more dynamic export prospects). Even where such strategic considerations are granted because of evident market failures, the theoretical approach tends to assign learning costs, externalities and support needs a relatively low value.

Finally, there is a belief that often tends to go with the above approach, but has no basis in the relevant economic theory. This is that, even when market failures are present, the risk and costs of government failure (in attempting to impose a corrective strategy that involves selective interventions) are greater. It is then better to suffer the consequences of market failure than those of poorly designed and badly implemented policies. In some interpretations of this political-economy argument, it is suggested that governments are intrinsically incapable of efficient selective interventions.

In conclusion, the general policy conclusions that derive from the approach just described are that there is little need for explicit technology-development policies in developing countries, or that this need is fairly limited and can be fully addressed with a functional (non-selective) approach. 'Getting prices right' is *necessary and sufficient* to promote the optimal pattern of industrial and technological development.[12]

The description of the technology-development process presented above may lead to more complex and differentiated policy conclusions. It suggests that active technology development is a vital ingredient of utilising industrial technologies. The effort needed may be risky, prolonged and variable across technologies and firms, and will depend on the level of development of the country concerned. There may be considerable externalities and need for support. An element of selectivity may be a necessary and desirable element of support policies, provided that certain safeguards are built in to cope with the real risks of government failure. The nature of the policies would vary by country and industry. The promotion of ITD is an important policy matter, and it does not always admit simple general solutions. These issues are further analysed below.

DETERMINANTS OF TECHNOLOGICAL DEVELOPMENT

The most basic pressure for an enterprise to undertake investments in TC is simply to get a new technology into production. This applies across all policy regimes and all types of firms. However, once production starts, different firms invest to different extents and with different degrees of success in achieving various possible technical objectives: reach full capacity, improve quality, adapt to local materials and demands, lower processing or energy costs, raise the productivity of equipment and workers, introduce new products or processes, find good suppliers and subcontractors, seek new information on technological progress elsewhere, and so on. What affects these decisions and outcomes?

The influences on TC acquisition can be grouped under five headings: the *incentive structure* facing firms; the availability of the right *quantity and quality of skills*; the availability of *technical information and support services*; *finance* for ITD investments; and the *technology policies* of the government. The complex interaction of these factors determines the willingness and ability of enterprises to develop capabilities. There is always a random element in the response of each individual firm, but there are sufficient commonalities that make an analysis of such determinants worthwhile.

Incentives

A large number of factors outside the firm may influence the decision to invest in TCs, from the size and growth of its markets and the pace of technical change to unexpected economic shocks. From the policy point of view, however, and leaving aside specific technology policies till later, the most important ones fall under three headings. First comes *macroeconomic policies*. It is evident that the rate and stability of growth of GNP, along with interest rates, price changes, exchange rates, fiscal and monetary policies and the availability of foreign exchange, will affect investments in TC just as they affect investments in physical capital. There is little to add here that is not already widely accepted among policy analysts.

The second set of incentives relates to the *trade regime*. The benefits of outward orientation to industrial development are again so well known that there is no need for extensive discussion here. Exposure to international markets seems to offer several stimuli to capability development. World class competition is a powerful catalyst to efforts to cut costs, improve quality, introduce new products and keep up with technical progress. Selling to world markets allows the realisation of scale economies. It also provides constant, and generally free, feedback for product design and process improvements. More broadly, it breeds dynamic managerial attitudes and flexible organisational structures. No amount of domestic competition can offer comparable benefits. These are all well known and widely accepted facts.[13]

Two additional points, however, need to be made. First, the main benefits of outward orientation for industrial efficiency have arisen not so much because countries have invested in activities in which they already had a static comparative advantage, but from *the efficiency with which they have utilised technologies* and from *the high rates of productivity growth they have enjoyed*. The gains in dynamic technical efficiency have, in other words, outweighed the gains in static allocative efficiency. This is the converse of the point made at the start of the chapter, on the sources of low productivity in developing countries, and it highlights the importance of boosting capabilities.

The second point is related to the first. *Outward orientation has not meant the absence of selective trade and other interventions* for the majority of NIEs. Most of them, especially the larger ones with a formidable market presence in a variety of complex manufactured exports, have combined strong incentives for exporting with a variety of interventions to protect, stimulate and guide their entry into difficult new activities.[14] These interventions have been selective (in the case of Korea very highly selective) as well as functional. The selectivity was designed precisely to overcome the barriers created by the high learning costs of entering demanding technologies. It has also been designed to exploit externalities and linkages between activities in what were deemed 'strategic' sectors for the development of the industrial base. Thus, they did not just promote particular technologies or products, but geared the choice to a larger plan based on their interrelationships.

The one NIE that did not practise selectivity, Hong Kong, started its industrialisation with very special skill and trading advantages, so that a lot of the 'learning' in its garment and textile activities had already been done elsewhere (in the experienced engineers and technicians that moved from the Chinese mainland). Moreover, because of its lack of promotion of more complex activities, it did not, over time, diversify and deepen its industrial structure (though it moved upmarket in the relatively simple technologies it utilised). In contrast, a smaller NIE, Singapore, practised selectivity not by offering import protection, but by directing the flow of industrial investments, by means of various incentives and pressures, into activities that it considered to be in its long-term comparative advantage. The learning costs were reduced by its investments in human capital (below), by relying heavily on foreign investors to bring 'ready made' technology (and the internal capital markets to finance the learning process), and by specialising in small segments of the technology rather than seeking to establish integrated facilities.

The lessons of this seem to be that it takes *time, investment and effort* to reach world levels of efficiency in complex technologies. The process is risky, uncertain and fraught with externalities. In developing countries the learning process is not understood, and firms often have to 'learn to learn'.[15] Individual firms will find it extremely difficult to bear the costs involved if they have been exposed to global competition from the start, and capital markets in developing countries are usually not prepared to risk such a process. There are thus valid infant-industry arguments to protect new industries, but they differ from the usual case for low and uniform protection. The duration and extent of protection cannot be uniform when different technologies have different learning costs and periods: the garment industry may be able to reach competence in three months, while automobile manufacture may, depending on the level of local content, take ten to fifteen years.[16]

The infant-industry case has now got a bad name because of the many horrors that have been perpetrated in its name. The developing world is littered with inefficient infants that have never matured. But this argues against unselective interventions without safeguards, and without complementary measures to address factor-market failures related to capability development. It is not *per se* an argument against selectivity, as the few very successful cases of selective intervention show. The practical lessons that emerge are:

- Industries to be promoted as infants must be capable of reaching world levels of efficiency in the foreseeable future, based on an evaluation of their skill and information requirements and the ability of the industrial and educational structure to meet those requirements.

- Only a few infants must be promoted at any given time, because intervention resources are limited. The interventions must be limited in time, taking account of the technologies involved. Industries that have matured must be faced with world competition. Performance has to be strictly monitored and enforced.[17]

- Protection takes away the incentive to invest in TC development. It can therefore be self-negating unless safeguarding measures are undertaken. In the experience of the NIEs, the most effective of these is the strong pressure to export a significant part of output fairly early in the process. This is why export orientation has such an impressive track record in promoting technical efficiency.

- Firms cannot become efficient on their own where there are strong linkages with other firms and new factor needs. Protection by itself cannot lead to competitiveness unless these linkages and factor-market needs are addressed – market failures have to be dealt with at their source. If the need is for a specific new type of engineering skill, the government has to ensure that the educational establishment creates that skill rather than protecting the producer. If gaining efficiency in one firm is based on the growth in efficiency of its suppliers, then again both related learning processes have to be promoted together. If learning requires large amounts of financing, the financial system has to be able to appraise and support such investments. And so on. In essence, protection has to be part of a larger, *coherent* package of measures to promote capability building.

The optimal incentive structure for ITD in the realm of trade policy in a world of uncertain learning, externalities and factor market failures is thus something approximating what was practised by the larger East Asian NIEs. This receives support both in economic theory and in the practical experience of the most successful industrialising countries in recent history.[18]

This analysis of incentives has been concerned with countries building up new industries. Most developing countries face a different problem: they already have substantial industrial structures that are inefficient and that need to be liberalised and reformed. These structures have been developed under uneconomic, non-selective and haphazard interventions, often with inadequate support from skill and institutional markets (see below). They have as a consequence invested relatively little in developing industrial capabilities, or have developed the wrong kinds of capabilities. Their capabilities have been directed more at 'making do' with available materials and adapting products to protected domestic markets than to reducing costs, improving quality and constantly introducing new products that have international markets.[19]

It has to be noted that many enterprises in import-substituting regimes have been able to master their technologies sufficiently to produce at or below world prices, and there are many complex and competitive industries there that would never have started had they not been given an initial period of protection. Many of the leading exporters in economies like Brazil, Mexico or Korea are heavy industries that 'learned' under import substitution. However, the quasi-permanent nature of protection and the lack of exposure to world markets in most inward-oriented economies have meant that their capabilities were not fully developed or exploited. There was a general tendency to remain technologically stagnant, and the overall cost of inefficiency and low productivity growth meant that the costs of such industrialisation were unjustifiable.

The most pressing policy need in these countries is therefore to restructure existing industries and to reallocate resources efficiently. This involves the liberalisation of trade and competition regimes so that efficient activities can grow and exploit their advantages in world markets. It also involves the shutting down of activities that cannot be made competitive within a reasonable period and with a reasonable level of restructuring investment. Most importantly, it involves the upgrading of activities that would be potentially efficient if their technological

and managerial capabilities were upgraded, and possibly some reha-
bilitation of equipment undertaken. The physical aspect of rehabilita-
tion is relatively easy (if the finance is available). It is the upgrading
of TCs that often takes time and effort. There is a process of 'unlearning'
and 'relearning' involved, in which many inherited attitudes, skills and
knowledge have to be replaced by new capabilities. This may be easy
in some cases, prolonged and difficult in others.

The design of the correct incentive structure to speed up the relearning
process faces many of the same issues as outlined above for infant
industries. Since the process involves time, investment and risk, an
immediate exposure to world competition may destroy activities that
are economical and viable. A carefully phased programme of liberalisation, with a coherent set of supporting measures on the skill and
technology fronts, could enable viable activities to upgrade to international standards.[20] It could thus save valuable industrial resources and
the 'good' parts of past capability development, which should also be
regarded as costly social investments. The phasing and sequencing of
structural adjustment should take into account the upgrading of TCs,
and the market failures that are inherent in this process.

The final point under incentives concerns *domestic competition policies*. The promotion of domestic competition, with free entry and exit
of firms, is an essential part of building healthy capabilities. This usually involves removing biases in favour of public enterprises or particular forms of ownership. It may also involve the privatisation of
public manufacturing firms and the restructuring of those that remain
in state hands, to ensure that they are able to respond to market pressures to invest in TC development. Artificial restraints on domestic
competition only retard TC acquisition, unless strong grounds for exception exist for natural monopolies or for highly scale-intensive activities (in which case appropriate regulations or import competition
have to be introduced to ensure that TC investments are made). There
is little new to add here.

Skills

There is general agreement on the need to strengthen the human-capital
base of developing countries to promote the growth of TCs. Human
capital is a broad term. In the industrial context, it includes the skills
imparted by the formal education and training system as well as those
created by on-the-job training and experience. Less advanced or more
specialised firms clearly need a smaller range than those in large-scale

and technologically complex industries. But even the simplest of technologies, if it is to be operated at world levels of efficiency, needs a range of worker, supervisory, maintenance, quality control and adaptive skills. At the lowest end of the technological spectrum, simple literacy and some vocational training, complemented by a few higher-level technical skills, may be sufficient to ensure adequate TCs (even garment assembly for export requires good supervisory, layout and maintenance skills). In more sophisticated technologies, the requirements are more diverse and the range of special skills wider.

Thus, while primary and secondary schooling provide the necessary base for building shop-floor capabilities in all activities, more advanced technical training becomes critical as the industrial structure develops. Since many forms of high-level skills are not fungible (a textile engineer cannot, for example, design electrical equipment), the range of specialisation grows with industrial complexity. At the more advanced levels, there are design and research needs which may call for doctorate holders (in the developed world, the traditional distinction between engineering and science is being breached as many new technologies call for science-based skills in production). The education system has to match the skill needs of the industrial structure, not just in the quantity of the people it graduates in different disciplines, but also in the quality of the training and the relevance of the curriculum. The ability to meet these needs has characterised the policies of the most successful NIEs.[21]

Formal pre-employment education is only one component of the skill-creation process. The other is on-the-job training and experience, and also further formal training of employees sponsored by the enterprises. The latter is a rapidly increasing activity among firms in developed countries. In Japan, in particular, where the lifetime-employment system encourages firms to invest in employee training, many firms spend more on training than on R&D. Japanese employee training expenditures in total are far larger than national expenditures on formal education. It is widely believed that this constitutes a major source of competitive strength for Japanese industry.

Most enterprises in developing countries invest little in upgrading the skills of their employees (East Asian NIEs like Korea and Singapore are striking exceptions). This may be for several reasons. First, managers may not be fully aware of the skill needs of the technologies they are using. They may be rooted in traditional ways of manufacturing and training (such as the apprenticeship system in Africa) that are unsuited to the needs of modern technologies or to the upgrading of existing technologies. Their own level of education may, in certain cases, make them averse to the further training of employees. These are symptoms of gaps in information markets and failures caused by attitudes and knowledge.

Secondly, managers may realise that employee skills are deficient, but may not be able to remedy this. They may not have access to trainers in-house. Overseas training may be out of the question. And there may be no local institutions, official or private, that can offer the right level and quality of training. These reflect failures in factor markets, caused by deficient educational facilities and a lack of responsiveness or information on the part of educators.

Finally, employers may be averse to investing in upgrading employee skills when they are unsure of being able to recoup the full benefits of their investments. Trained employees may leave the firm, taking the benefits of training to competitors – a classic externality-caused market failure that deters socially desirable investments in human capital (this is also often true of developed countries, of course, which is why the lifetime-employment system in Japan was noted above).

The risk of market failures in skill creation is generally accepted, even by institutions such as the World Bank which are essentially averse to all kinds of intervention.[22] These interventions are assumed to be 'functional', simply strengthening factor markets without attempting to support selected activities over others. This assumption is not always justified. Many forms of educational investments *are* non-selective, providing a general base of skills for all activities. This is so particularly for schooling and non-technical higher education. By contrast, more specialised forms of tertiary education may be highly industry-specific. As noted above, the specificity of skills increases with the development of the industrial structure. If the education market cannot fully anticipate the skill needs of industries in time to provide an adequate base of engineers and technicians, the interventions needed have to be selective.

The selectivity of educational policies increases with the role that the government plays in determining the direction of industrial growth. If the government has an active industrial strategy, pushing the economy into more complex areas of activity, it has to select the skills that the education system has to create accordingly. Selectivity in one sphere necessarily calls for selectivity in all related, supporting spheres. This coherence is what marked the success of interventions in NIEs like Korea, Taiwan and Singapore.

Technical information and support services

Most of the technological activity that firms undertake to cope with production problems, cost reduction, diversification and so on is an in-house affair. However, many firms in developing countries may not know that they have to undertake deliberate technological effort to resolve those problems: they have to be taught to invest in TCs. When they do undertake the effort, they often draw on information and help from outside. There are many problems that firms may not have the specialised skills, equipment or resources to solve by themselves. This is where the role of persuasion, information and technical support becomes significant.

Technical information and support may exist in many forms. A lot of information is available free: from journals, contacts with capital-goods suppliers and buyers of export products, visits to fairs, plants and conferences, interactions with subcontractors and other suppliers, and so on. More complex or closely held information is available commercially from consultants, more advanced firms (on licence), or as part of a package of direct investment.

Apart from these, there is a range of information and support services that are not provided by the market in any country, especially by markets in developing countries. There are several technological functions that have 'public-goods' features, whose rewards are difficult to appropriate by private firms and so suffer from market failure. These include the encouragement of technological activity in general (overcoming risk aversion and the 'learning to learn' barrier); the development of special research skills; the setting of industrial standards and the promotion of quality awareness: the provision of metrology (industrial measurement and calibration) services; the undertaking of contract research, testing or information search for firms that lack the facilities or skills; other extension services for small enterprises; and the undertaking and coordination of basic (pre-commercial) research activities.

The provision of these services then has to be undertaken as an infrastructural service, or as a cooperative activity by the enterprises concerned. Cooperative provision of some forms of training, research, design, quality assurance and information collection does exist in many countries. However, the development of an efficient cooperative technology support system calls for a fairly advanced industrial structure, and its evolution may take a long time. As in Germany in the late nineteenth century, the government may help to introduce cooperative solutions with financial and other assistance. The remainder falls on the government to provide as an infrastructural service. Most developing countries tend to treat practically all these tasks as a public responsibility, under the banner of the 'science and technology infrastructure'.

The public infrastructure for science and technology (S&T) may comprise a large range of institutions, from universities and research financing bodies to standards institutes and R&D institutions. The efficacy of this infrastructure has been variable. In many developing countries S&T institutions have been inadequately staffed and poorly equipped and funded. In some a lot of money has been invested, but without a proper linkage between the technological needs of industry and the research or promotion work of the institutions. In the poorest and least developed countries, there is no research conducted worth the name, and there is little effective extension work to support small enterprises. In the more advanced countries, the technical extension is somewhat better and some high-quality research is carried out, but most of the latter is divorced from industrial technology. There are exceptions, of course, but on the whole the infrastructure has not served the functions it was set up for.

This is not to say that there is no need for S&T institutions. The promotion of ITD does require infrastructural support, especially as the industrial structure develops and as information flows and technical specialisation become more important. Export promotion requires an active standards institute, not merely to inform producers of the required standards and to vet product quality, but to play an active role in helping industry to achieve those standards – a technically demanding task. The observance of standards and the carrying out of research calls for precise measurements and calibration, for which a metrology office becomes essential. Contract research is needed by smaller firms, even by larger ones for specific problems, as they try to upgrade their technologies and introduce new products. Access to basic scientific knowledge can help firms to do applied research in some areas. And so on.

The need for information and technical support grows with the level of TC development, but some needs exist even at the starting levels. The weakness of the support structure is itself a symptom of the low levels of skill and TC development, but the strengthening of the structure can greatly help the technological development process.

Finance

Capital-market failures are commonly, and correctly, regarded as one of the main barriers to technology development. Many of the failures in developing countries are policy induced. Financial repression, arbitrary allocation of credit and other interventions interfere with efficient resource allocation by the banking system. Apart from these, however, there are some inherent features of the financial system that may make it unsuited to the needs of TC investments.

There is an important distinction to be made here. A large part of TC development, especially at the lower levels, is not the result of formal research activity but of problem solving in the course of production. The financing of this form of TC activity does not require special instruments apart from those that normally apply to industry. The rest of TC development is based on more formal technological effort, involving longer-term search and experimentation, and the introduction of new processes or products that face considerable market risk. This form of technological activity, more significant at advanced levels of industrialisation, calls for different financing mechanisms, or at least for more imaginative approaches from conventional sources of finance.

A variety of financing mechanisms exist for technology development of the latter sort. Some involve loans on special terms or at concessional rates. Some involve equity participation by the lender. Some are a mixture of loans and equity. Many of these forms appear themselves as financial markets perceive the need for new instruments. In developing countries, however, this responsiveness is often lacking in the financial system (perhaps partly as a result of misguided controls and interventions). The first requirement is to remove inefficient interventions. There may then be a role for the government in promoting appropriate financial mechanisms and institutions for ITD.

Technology Policies

The final determinant of TC investments is the set of explicit technology policies adopted by the government. Many of the indirect policy influences have been noted already. Direct policies include fiscal or other incentives for technological activity (by individual firms or groups of firms together), other forms of support (such as procurement policies), regulations on imports of foreign technologies and direct foreign investments, and the direct targeting of specific technologies for research by the public sector (perhaps in collaboration with private enterprises). Of these, the ones of greatest practical significance seem to be the control of technology imports and direct foreign investment, and 'mission oriented' R&D strategies. Fiscal incentives are probably marginal in stimulating real investments in technological activity, though efforts to coordinate inter-firm R&D are sometimes useful in promoting basic (pre-commercial) advance in frontier technologies. Procurement preferences for goods based on local R&D are protective instruments. If used sparingly and carefully, they can stimulate TD. If not they can lead to the featherbedding of inefficient producers.

The control of technology imports has been common among developing countries. The international technology market is one of the most imperfect of markets. There are large elements of monopoly or oligopoly. Information on sources and 'products' is often fragmented and costly to collect. It is difficult for the buyer to evaluate the real value of the 'product' or the going market price. The product itself has some public-goods properties, since it is relatively expensive to create but cheap to disseminate, and the initial stock does not diminish with the sale. All these imperfections have led governments in many developing countries to impose all sorts of controls on technology transfers, to lower the cost and improve their enterprises' bargaining positions.

The results have fallen short of expectations. Rigid and cumbersome regulations on the content and terms of technology transfer have tended to reduce the quantity and quality of inflows, to the detriment of the buying country. The growth of local TCs has been hampered rather than helped by over-restrictive policies. The relationship between the import and local generation of technology is sometimes competitive, but it is also often complementary. Firms should generally be left to make their own decisions on where and what to buy in the technology market – what is critical is to get the larger policy and incentive framework right.

Similar considerations apply to foreign direct investments. Foreign investment flows are dominated by large, oligopolistic multinational companies that exercise considerable market power in their respective industries. Until recently, their entry was regarded with suspicion and hostility by many developing countries, though they were aware that the leading multinationals were the main sources of industrial innovation. This attitude cost many countries the direct access to modern technologies, and to the skills, finance and marketing that went into making those technologies operational.

Two points need, however, to be noted about technology development and foreign direct investment. First, though investors can provide access to the most modern technologies, they transfer only those technologies that the host country, with its skills, capabilities, supplier base and infrastructure, can absorb. Unless it is highly protected or subject to major distortions, a country with poor indigenous capabilities will only get simple technologies. While the investor will try to en-sure that these are efficiently implemented, the further upgrading of the industrial structure will depend upon the effort the country puts into building up its skills, suppliers and so on. Ultimately, the foreign investor cannot do the necessary 'homework' needed to support capa-bility development. All it can do is to create islands of static efficiency. There is no substitute for indigenous capability development effort, and technological transformation cannot be a passive process based on open doors to foreign investors.

Secondly, foreign investment is an effective means of transferring the results of innovation, but *not necessarily the innovation capability itself*. Economies of scale, agglomeration and access to advanced skills and S&T infrastructures lead multinational firms to centralise their innovation work in a few developed countries, with the main base generally in their country of origin. The R&D that is conducted in a few developing countries is small in quantity, and tends to be geared to adapting local raw materials and products. It is rarely part of the core innovative process in the industry concerned.

This is of little practical significance to the majority of developing countries that cannot possibly support design and innovation. All they need is to import operational technology (the 'know-how') efficiently. However, the more advanced countries, the NIEs and near-NIEs, need to develop deeper understanding of the nature of the technologies they use (the 'know-why'), not necessarily to push back world frontiers but to be able to master the technologies fully and so to adapt and improve on them and diversify their industries. They need, in other words, to deepen their technological capabilities in order to undertake the more advanced design and development tasks. As the industrial structure deep-ens, the growth of more advanced TCs enables countries to tackle more complex activities, even if they continue to rely on innovations made elsewhere. The Japanese experience of developing indigenous capabilities while importing innovations shows how the process of technological deepening can have widespread externalities and dynamic effects.

In many technologies, however, it is now not feasible for develop-ing countries to 'go it alone' technologically. In these the entry of foreign investors should be encouraged, to boost technological capa-bilities. In others, the optimal choice of technology transfer form should depend on the absorptive and innovative capacities of local firms and the willingness of foreign investors to invest in deepening local capa-bilities. In some cases, the best policy may be to promote local learn-ing by insisting on majority or full local ownership, backed by measures to ensure that the enterprises invest in TCs and achieve international competitiveness quickly (the Korean approach). In others, it may be possible to induce multinationals to shift some of their design and development work to developing-country affiliates (the Singapore solution). And in some it may be possible for the government itself to take the coordinating role and promote local learning in joint ventures with multinationals (the Taiwanese approach in semiconductor manufacture). All these approaches involve widespread capability-building measures by the country, with investments in education, the local S&T infrastructure, and strong export orientation.

Ultimately, there is no single optimal policy for developing countries on foreign investment. Past policies have generally been too restrictive and interventionist, and retarded the technological upgrading of developing countries. On the other hand, simply opening doors to foreign investors is not a solution to the problems on technological development. A wholesale dependence on foreign investors for all difficult technological work can curtail the learning process, and may channel industrialisation into shallower, less dynamic paths. The experience of the larger East Asian NIEs again shows that the deliberate policy to encourage domestic 'know-why', with selective entry of foreign investors, can yield significant dividends for industrial development at a certain stage of the process.

This concludes the discussion of the determinants of TC development. As noted, it is the *interaction* of the various factors that decides the final outcome, not any particular set of factors by itself. Just getting the incentive structure right would be of little use if the capacity to respond to the incentives were deficient. Investing in education and technology support systems would be worthless if the incentive structure misguided the allocative decisions of firms. In all aspects of TC development, there may be a constructive role for government policies. This is considered now.

MARKET FAILURES AND POLICY NEEDS FOR ITD

The policy needs of industrial technology development are shown by two things: the nature of market failures for each determinant, and the capability of the government concerned to devise solutions that are better than market outcomes. This section deals with the first. The next section deals with the second. The present section can be fairly

Table 2.1   Determinants of technology development, market failures and policy remedies

Determinants	Market Failures	Policy Remedies
INCENTIVES: Macroeconomic policies. Free Trade. Domestic competition.	Not applicable. Externalities, dynamic learning, information gaps, risk, capital-mkt failures, inherited attitudes and capabilities. Market power, scale economies, complementarities.	Infant-industry protection for new activities with difficult learning (very selective, monitored, limited in duration, with safeguards, integration with skills & institutional development). For reform of existing industry, phased liberalisation, taking account of relearning costs, integrated with other support measures. Ensure competition, regulate monopoly, exploit complementarities.
SKILLS: Worker and supervisory. Technical. Production engineering. Design, research, development. Scientific. Managerial, organisational, marketing.	In the formal educ. system, investments suffer from lumpiness, lack of supply (teachers and facilities), imperfect foresight, lack of information. Quality control and curriculum content suffer from opportunism, information gaps. Investments by firms in training suffer from externalities (lack of appropriability), lack of knowledge of benefits of training, risk aversion, capital-market failures.	Govt support of primary and secondary schooling, and of higher-level educ. where necessary. Provision of special training, control of quality and content. Incentives, subsidies for in-firm training and/or setting up cooperative facilities. Support for some foreign training, import of foreign trainers. (Cost of higher-educ. needs to be carefully controlled. Skill profile has to be relevant to industrial needs.)
INFORMATION & TECHNICAL SUPPORT: Knowledge of need for TD effort. Knowledge of kind of effort to make. Access to information from other firms, institutions, universities, etc. Standards, metrology, testing facilities. Technical extension services. Contract research, design, training. Information services on tech., sources, trends, etc. Basic research support.	Information gaps and fragmented information markets; 'learning to learn' delays; lumpiness of facilities; externalities and lack of appropriability; skill gaps, risk aversion, absence of technological intermediation.	Information and persuasion of firms of the need for and value of tech. activity. Strengthening of intellectual property rights. Provision of infrastructural services; setting up R&D institution (ensuring that they are linked to enterprises). Technology extension services for small firms. Information services on sources of technology. Support of cooperative R&D by industries; mission-oriented R&D support (this has to be economic and geared to real industrial needs).

FINANCE FOR ITD: Availability of finance at appropriate rates and in sufficient quantity for R&D or the commercialisation of **innovations**. Equity-sharing finance for innovators. Special finance for small and medium enterprises.	Capital-market failures from asymmetric or missing information, moral hazard, cost of evaluation or enforcement of ITD loans; risk aversion or over-conservative policies by fin. intermediaries of relevant financial intermediation skills.	Creating technology-financing capabilities in banks, with training, subsidies (to start with only); special financial provision for ITD efforts that link R&D institutes with enterprises; financial instruments for SMIs; venture capital and other schemes to provide special instruments for risk sharing.
TECHNOLOGY POLICIES: Technology imports, FDI, promotion of local R&D and absorption of foreign technologies, other interventions to strengthen ITD.	Insufficient investment in local R&D due to factors above. Transfer of tech.: international technology-market imperfections, monopoly, lack of or asymmetric information, passive dependence on imported technology, plus other failures above that deter ITD.	Fiscal and other incentives for R&D; procurement of products of local innovations (very selective and careful use); information service on sources of tech. (but minimum interference with tech. transactions); selective control of FDI and negotiation to ensure local 'know-why' development (only relevant for advanced countries and some activities).

brief, because the relevant market failures have been noted in the earlier discussion. It will simply synthesise the points made there.

Table 2.1 sets out in summary form the main determinants of ITD, the nature of possible market failures (not those resulting from government interventions), and the range of possible policy solutions. The table does not cover all possible market failures that can affect ITD (for instance, it does not go into the promotion of inter-industry or inter-firm linkages). It is intended to provide an impression of the main issues that arise in this area, including those arising from the structural adjustment process.

It may be reiterated that not all market failures call for interventions. Markets may themselves throw up solutions in the form of new contractual relationships (inter-firm linkages), institutions (industry association-sponsored research or extension bodies) and enterprises that develop marketable commodities from the needs of industry (technology intermediaries, specialised consultants). The need for interventions arises when these solutions do not appear, or need encouragement to appear. In the latter case, interventions may be only supportive of market-based solutions. It is not possible to say *a priori* how many market failures need to be dealt with in this way, and how many require permanent government presence.

If physical infrastructure is any guide, many services that were traditionally thought to be in the state preserve may be privatised efficiently, if proper regulations are instituted to make sure that natural monopolies are under constant pressure to cut costs, improve quality and innovate. In the case of S&T infrastructure, many services can be similarly placed in the private sector, and operated on a commercial basis. This does require a fairly well developed private sector, and a well established demand for technology services. In the early stages of technology development, where the government may have to do a lot of hand holding, it may be best to keep these in the public sector, or at least to subsidise the private provision of these services.

GOVERNMENT FAILURE

The risk of government failure has to be faced whenever policy solutions are recommended for market deficiencies. The widespread intervention of well-meaning governments has been the cause of enormous waste and inefficiency in the developing world. This has become so evident that the current reaction is to treat all governments as basically incapable of successful intervention. However, governments do not succeed and fail absolutely. As with markets, there are *degrees* of failure and success. And, as with markets, governments can improve with time and effort. For some tasks, there may be no alternative to state provision. With others, a reasoned judgement has to be made on the costs and benefits of intervention versus market failure.[23]

It is widely accepted that the provision of basic education and infrastructure services in relation to ITD should be in government hands in most developing countries. These are the 'market-friendly interventions' that the World Bank now favours. However incompetent governments are, there is no realistic alternative to their participation in these areas. The best that can be done is to strengthen their capabilities and improve their performance, making them as responsive to market forces as possible. The more serious problems start when it comes to more selective forms of intervention.

These selective policies may be of several types: the creation of particular types of skill, the setting up of institutions to promote particular 'strategic' technologies, the financing of 'mission-oriented' research, the granting of infant-industry protection or subsidies, the channelling of local or foreign investments into particular activities, or negotiating with and regulating international investment and technology transfers to achieve technological objectives. These are all policies that require enormous skill, information and discipline on the part of the government. They are prone to rent-seeking behaviour and pressure groups. They can be very costly if they are wrongly designed and implemented.

The experience of the most dynamic industrialisers in the developing world suggests that their selective interventions determined the nature and success of their industrial development, and that such interventions were well implemented. The real issue is whether the conditions under which their governments operated are replicable in other countries. Do the governments of typical developing countries, in other words, have the clarity of economic objectives, information, design skills, implementation/monitoring capabilities and the political will to carry out selective policies in the way that the larger NIEs of East Asia did?

The answer must be that most do not. The level of selectivity exercised, say, by the Korean government was so detailed and pervasive that few other countries can hope to emulate it with a comparable degree of success. It also had many advantages that may be unique: the homogeneity of the population, the relatively favourable income distribution and the mobilisation of nationalist sentiment in response to an external threat. It started with a strong base of human capital, which the government was able to enhance dramatically, partly because of local traditions. It was driven by the single-minded pursuit of particular economic objectives. It had direct economic involvement at the highest political level, close government collaboration with private business, a dedicated and well-trained civil service, the ability to monitor performance and to penalise poor performers. Its design was closely influenced by the experience of neighbouring Japan, with which there were cultural similarities and technological links. There can be few countries that have this combination of characteristics.

This means that the complexity and range of Korean interventions may not be replicable. It does not necessarily mean, however, that no selective interventions are possible. There are valuable lessons to be learned from Korea, if they are adapted to the economic conditions and political realities of each country. For example, the need for promotion of infant-industry learning is universal in developing countries. Faced with 'ready made' capabilities in advanced countries, they cannot absorb the costs and risks of mastering difficult new technologies unless they are assisted in factor markets and offered some security of markets or profits. However, this promotion need not be as detailed and specific as in Korea, where the government intervened at the level of industry, product, technology and firm. More general levels of protection, at the broad subsectoral level, may be suited to countries that have limited administrative capabilities. Market forces can then be left to sort out the best enterprises and technologies.

The most useful lessons of the Korean experience are perhaps that many of the potential costs of protection can be overcome by instituting safeguards (early entry into export markets) and integrating the incentive interventions with interventions on the supply side. Most developing countries have granted protection without ensuring that their enterprises had access to the new skills and information they needed to become competitive. Unlike the Koreans, they were not *truly* selective in planning industrial strategy.

The correct balance of interventions and market forces cannot be decided on *a priori* grounds. Both governments and markets fail. Both can be improved with effort. There are some governments today that are capable of mounting effective selective interventions at a fairly complex level. There are many that are best confined to functional policies, or the most general levels of selectivity. The judgement ultimately has to be a pragmatic one. *General statements on the virtues of markets versus governments are suspect, and may be economically harmful.*

CONCLUSIONS

The analysis of technological capabilities is essential to an understanding of the industrial-development process. The usual simplifying assumptions that there are no learning costs in using industrial technologies, and that efficient production can be launched merely in response to 'right' prices, do not do justice to a complex reality. They may lead to misleading policy recommendations. The discussion in this chapter suggests that the promotion of capability development is a vital part of the strategy of industrial development. It also suggests that, in the presence of market failures, active government involvement is required to ensure capability development.

This involvement must be based first on the removal of an accretion of many irrational and inefficient interventions that many developing-country governments have undertaken since the current industrialisation drive began. It then has to address several determinants of TC activity: the incentive framework, the supply of human capital, the supporting technology infrastructure, finance for ITD, and access to foreign technologies.

The best incentive framework for ITD is one which provides constant competition to enterprises in a stable macroeconomic setting. Full exposure to world competition has, however, to be tempered by the fact that a new entrant has to incur the costs and risks of gaining technological mastery, when its competitors in more advanced countries have already gone through the learning process. Depending on the extent of the learning costs and the efficiency of the relevant factor markets and supporting institutions, there is a case for infant-industry protection. But protection itself reduces the incentive to invest in capability building. It has to be carefully designed, sparingly granted, strictly monitored and offset by measures to force firms to aim for world standards of efficiency.

The supply of human capital, technical support services, foreign technology, S&T infrastructure and finance requires intervention because the relevant markets in developing countries generally suffer from a number of deficiencies. A sound ITD-promoting strategy must address each of these needs. Needless to say, these interventions must be integrated with interventions in the incentive framework, so that activities being promoted are not penalised by the lack of production factors and information. Intervention resources are scarce, and their most effective use calls for selectivity and coherence.

There are many possible levels and patterns of ITD support. The greater the degree of selectivity exercised, the greater the potential benefits if the strategy works, but the larger also the risk of government failure (and of heavy ensuing economic costs). A few governments have managed to intervene selectively with great success, and have produced industrial growth rates perhaps unmatched in recent economic history. Most have not. This means that government intervention capabilities have to be assessed in drawing up ITD strategies. It also means that these capabilities have to be developed to the extent possible. Realistically, however, optimal ITD strategies will for the time being involve relatively low levels of selectivity in most developing countries. What these levels are can only be decided on a case-by-case basis.

Much of ITD policy must focus on functional interventions to strengthen the technology infrastructure, improve its relations with enterprises, help small and medium-sized firms with their special information and support needs, augment the supply of finance for technology investments, and address the most pressing skill needs for efficient industrial operation and growth. The correct balance between these must vary by country and level of industrial development, but every developing country can benefit from well-designed ITD support. What is appropriate in, say, Ghana will be very different from what is appropriate in India, but the underlying principles are similar. In fact, one of the major challenges in this field may be to devise ITD strategies for the least industrialised economies: their technological needs are more pressing, if more rudimentary, than those of the advanced ones.

REFERENCES

Amsden, A. (1989) *Asia's New Giant: South Korea and Late Industrialisation* (New York: Oxford University Press).

Bell, M., Ross-Larson, B. and Westphal, L. E. (1984) 'Assessing the Performance of Infant Industries', *Journal of Development Economics*, 16(1): 101–28.

Chandler, A. D. (1992) 'Organizational Capabilities and the Economic History of the Industrial Enterprise', *Journal of Economic Perspectives*, 6(3): 79–100.

Dahlman, C. J., Ross-Larson, B. and Westphal, L. E. (1987) 'Managing Technological Development: Lessons from newly industrialising countries', *World Development*, 15(6): 759–75.

Enos, J. (1992), *The Creation of Technological Capabilities in Developing Countries* (London: Pinter).

Fanelli, J. M. and Frenkel, R. (1993) 'On Gradualism, Shock Treatment and Sequencing', in UNCTAD, *International Monetary and Financial Issues for the 1990s* (New York: United Nations).

Jacobsson, S. (1993) 'The Length of the Learning Period: Evidence from the Korean engineering industry', *World Development*, 21(3): 407–20.

Katz, J. (1984) 'Domestic Technological Innovation and Dynamic Comparative Advantage: Further reflections on a comparative case study program', *Journal of Development Economics*, 16(1): 13–38.

Katz, J. M. (ed.) (1987) *Technology Generation in Latin American Manufacturing Industries* (London: Macmillan).

Lall, S. (1987) *Learning to Industrialize* (London: Macmillan).

Lall, S. (1990) *Building Industrial Competitiveness in Developing Countries* (Paris: OECD Development Centre).

Lall, S. (1992) 'Technological Capabilities and Industrialization', *World Development*, 20(2): 165–86.

Lall, S. and associates (1992) *World Bank Support for Industrialization in Korea, India and Indonesia* (Washington, DC: World Bank Operations Evaluation Department).

Nelson, R. R. (1981) 'Research on Productivity Growth and Productivity Differences: Dead ends or new departures?' *Journal of Economic Literature*, 19(3): 1029–64.

Nelson, R. R. (1987) 'Innovation and Economic Development: Theoretical retrospect and prospect', in J. Katz (ed.), *Technology Generation in Latin American Manufacturing Industries* (London: Macmillan), pp. 78–93.

Nelson, R. R. and Winter, S. J. (1977) 'In Search of Useful Theory of Innovation', *Research Policy*, 6(1): 36–76.

Nelson, R. R. and Winter, S. J. (1982) *An Evolutionary Theory of Economic Change* (Cambridge, MA: Harvard University Press).

Pack, H. (1992) 'Learning and Productivity Change in Developing Countries', in G. K. Helleiner (ed.), *Trade Policy, Industrialization and Development*, (Oxford: Clarendon Press), pp. 21–45.

Pack, H. (1988) 'Industrialization and Trade', in H. B. Chenery and T. N. Srinivasan (eds), *Handbook of Development Economics*, Volume I (Amsterdam: North-Holland), pp. 333–79.

Pack, H. and Westphal, L. E. (1986) 'Industrial Strategy and Technological Change: Theory versus reality', *Journal of Development Economics*, 22(1): 87–128.

Porter, M. (1990) *The Competitive Advantage of Nations* (New York: Free Press).

Rodrik, D. (1992) 'Closing the Productivity Gap: Does trade liberalization really help?', *Journal of Economic Perspectives* 6(1): 87–105.

Shapiro, H. and Taylor, L. (1990) 'The State and Industrial Strategy', *World Development*, 18(6): 861–78.

Stiglitz, J. E. (1987) 'Learning to Learn, Localized Learning and Technological Progress', in P. Dasgupta and P. Stoneman (eds), *Economic Policy and Technological Development* (Cambridge: Cambridge University Press), pp. 125–55.

Stiglitz, J. E. (1989) 'Markets, Market Failures and Development', *American Economic Review Papers and Proceedings*, 79(2): 197–202.

Teitel, S. (1984) 'Technology Creation in Semi-Industrial Economies', *Journal of Development Economics*, 16(1): 39–61.
Wade, R. (1990) *Governing the Market: Economic Theory and the Role of Government in East Asian Industrialization* (Princeton: Princeton University Press).
World Bank (1991) *World Development Report 1991* (Washington, DC: World Bank).

3 Technology Development Policies: Lessons from Asia[1]

INTRODUCTION

This chapter describes the policies of some Asian countries for industrial technology development (ITD). It is not a comprehensive review of each country's technology policy, but selected descriptions of its important elements. It illustrates the nature of technology markets and problems in developing countries, and the policies that can be adopted to improve them. More interestingly, it shows that while they had similar approaches to some technological problems, the NIEs dealt with several important issues very differently. To clarify these differences, this chapter draws a distinction between different strategies to deal with failures in technology markets.

In economic theory, the case for intervention is measured by the costs and benefits of dealing with market failures. Technology markets are notoriously prone to failure. These failures are not identical across countries; they differ according to the level of development, the industrial structure and the initial base of skills and institutions.[2] Perhaps more importantly, they also differ according to the different *perceptions of governments of what constitutes 'market failure'* in technology development – interventions can go beyond correcting for existing market deficiencies (in a static sense)[3] to changing the parameters within which markets function (in a dynamic sense), by creating new factor endowments, institutions and market structures. It is difficult to describe the latter set of interventions as remedying 'market failures' in the neoclassical sense, since this defines failures with reference to a competitive equilibrium. In principle, markets can clear within a given set of endowments and parameters, even if these occur at low levels of income and growth. The conventional market-failure approach has little to say on changing those endowments and raising the economy beyond 'low-level equilibrium'.

Yet changing endowments and developing new market structures is what development policy is all about. Most governments seek to dynamise

59

their economic growth, create new sources of comparative advantage, deepen the industrial structures, and expand their base of technological capabilities. In addition, they often aim to develop specific industries and enter into new groups of activities that are believed to be more conducive to growth, competitiveness and technological progress than others.[4] In other words, industrial and technology policies can be selective rather than just 'market-friendly', as discussed elsewhere in this book.

ITD policies can thus be divided into two broad groups: those that address market failures in the conventional sense (deviations from static efficient markets), and those that seek to change basic endowments and parameters, guided by a strategy of longer-term development. The former can be described as *strategic*, the latter as *static*. In Asia, technology policies had many common static elements, mainly dealing with generic market failures that affect technology development in all developing, and most developed, countries. They also had striking differences in their strategic policies, reflecting different ideologies and political economies. Both are of interest to the developing world.

PROMOTING INDUSTRIAL RESEARCH AND DEVELOPMENT

While it is common to regard the stimulation of industrial research and development (R&D) as the main, or even the sole, aim of technology policy, this is (as the previous chapter argued) only one component of measures to raise technological competence, especially at low levels of industrial development. Nevertheless, formal R&D by industrial firms assumes increasing significance with industrial maturity, even in countries that have not reached the 'frontiers' of innovation. As more complex technologies are imported and deployed, R&D becomes important as a means of keeping track of and absorbing new technologies. A growing base of R&D capabilities also permits better and faster diffusion of new technologies within the economy, lowers the cost of technology transfer, and captures more of the spillover benefits created by the operation of foreign firms. Most importantly, it permits the industrial sector as a whole greater flexibility and diversification of industrial activity, and allows it greater autonomy by creating a 'technology culture'. The significance of a growing R&D base is widely acknowledged in the developed world; it is important to realise that this is almost as important for middle-level industrialising countries.

There can be various market failures in stimulating the growth of a

Table 3.1   R&D (% GDP)

Country	Year	Total R&D	R&D by enterprises
Japan	1988	2.8	1.9
Korea	1992	2.1	1.7
Taiwan	1993	1.8	0.9
Singapore	1992	1.0	0.6
Hong Kong	1995	0.1	N/A
Malaysia	1992	0.37	0.17
Thailand	1987	0.2	0.03
India	1990	1.0	0.2

Sources: UNESCO, *Statistical Yearbook*, various. Data on Taiwan from *Statistical Data Book 1994*, Government of Taiwan; and on Hong Kong from the *Far Eastern Economic Review*, 21 December 1995, p. 51.

'technology culture' in an industrialising economy.[5] There are well-known difficulties of appropriating fully the returns to private R&D; in newly industrialising countries the problems are compounded by the extra cost and risk of developing local research capabilities when technology can be imported from more advanced countries. There is a difficult choice to be made between importing 'ready made' technologies and developing the capabilities to adapt, modify and improve upon them. Clearly, too much stress on one or the other can be uneconomical. A heavy dependence on technology imports can be costly and may lead to a lack of technological dynamism; an over-emphasis on indigenous technology creation can lead to costly efforts to 'reinvent the wheel'. Policies to stimulate local R&D clearly fall in the category of strategic choices – there is no clear market failure involved in remaining highly dependent on foreign technology.

Developing countries have very different propensities to conduct their own R&D. Table 3.1 provides data on R&D propensities, in total and by productive enterprises, in the NIEs, Japan and India. Korea is the most successful of the NIEs in stimulating industrial R&D, and is approaching Japanese levels in terms of percentage of GDP (though, of course, the absolute amounts are much lower). In fact, Korean R&D spending by industry is now higher as a proportion of GDP than all OECD countries with the exception of the few technological leaders. Taiwan comes next, but with half of total R&D performed by the government. Singapore and India have about the same percentage of GDP in R&D, but in Singapore 60 per cent comes from industry, while in India the figure is only 20 per cent. Hong Kong spends negligible amounts on R&D. The new NIEs have much lower R&D propensities, though it is interesting that MNC affiliates in Malaysia are following the Singapore example and increasing local R&D: some 45 per cent of total R&D in Malaysia is performed by the industrial sector, led by foreign firms. Thailand, by contrast, has very low industrial R&D.

Korea

The Korean government combined selective import-substitution with forceful export promotion, protecting and subsidising targeted industries that were to form its future export advantage.[6] In order to enter heavy industry, promote local R&D capabilities and establish an international image for its exports, the government promoted the growth of giant local private firms, the *chaebol*, to spearhead its industrialisation drive. Korean industry built up an impressive R&D capability by drawing extensively on foreign technology in forms that promoted local control. Thus, it was one of the largest importers of capital goods in the developing world, and encouraged its firms to obtain the latest equipment (except when it was promoting particular domestic products) and technology. It encouraged the hiring of foreign experts, and the flow (often informal) of engineers from Japan to help resolve technical problems.

FDI was allowed only where considered necessary, and the government sought to keep control firmly in local hands. Foreign majority ownership was not permitted unless it was a condition of having access to closely-held technologies, or to promote exports in internationally integrated activities. The government intervened in major technology contracts to strengthen domestic buyers, and sought to maximise the participation of local consultants in engineering contracts to develop basic process capabilities. In 1973, it enacted the Engineering Service Promotion Law to protect and strengthen the domestic engineering-services sector, and the Law for the Development of Specially Designated Research Institutes to provide legal, financial and tax incentives for private and public institutes in selected technological activities.

Technological effort in Korea was supported by the government in several ways. Private R&D was directly promoted by a number of incentives and other forms of assistance. *Incentive schemes* included tax exempt TDR (Technology Development Reserve) funds, tax credits for R&D expenditures as well as for upgrading human capital related to research and setting up industry research institutes, accelerated depreciation for investments in R&D facilities and a tax exemption for 10 per cent of the cost of relevant equipment, reduced import duties for imported research equipment, and a reduced excise tax for technology-intensive products. The KTAC (Korea Technology Advancement Corporation) helped firms to commercialise research results; a 6 per cent tax credit or special accelerated depreciation provided further incentives.

The import of technology was promoted by tax incentives: transfer costs of patent rights and technology import fees were tax deductible; income from technology consulting was tax-exempt; and foreign engineers were exempt from income tax. In addition, the government gave *grants* and *long-term low-interest loans* to participants in 'National Projects', which gave tax privileges and official funds to private and government R&D institutes to carry out these projects. Technology finance was provided by the Korea Technology Development Corporation (below).

However, the main stimulus to the tremendous growth of industrial R&D came less from the specific incentives to R&D than from the overall incentive regime that created large firms, gave them a protected market to master complex technologies, minimised reliance on FDI, and forced them into international markets where competition ensured that they would have to invest in their own research capabilities. This is why, for instance, Korea has 35 times higher R&D by industry as a proportion of GDP than Mexico (with roughly the same size of manufacturing value added), an economy that has remained highly dependent on technology imports (Najmabadi and Lall, 1995).

Taiwan

The history of Taiwanese R&D growth has some similarities to Korea, but there are also marked differences because of their different political economies and industrial structures: the Taiwanese government always had a more distant relationship with industry and never promoted the growth of large private conglomerates like the *chaebol*. It started to address the development of local R&D capabilities fairly early, in the late 1950s, when its growing trade dependence reinforced the need to enhance local innovative effort to upgrade and diversify its exports.[7] A Science and Technology Programme was started in 1979, targeting energy, production automation, information-science and materials-science technologies for development. In 1982, biotechnology, electro-optics, hepatitis control and food technology were added to this list. In 1986, the current S&T Development Plan (1986–95) was launched, continuing the targeting of strategic areas of technology. It set a target for total R&D of 2 per cent of GDP for 1995; by 1993, the country had reached 1.8 per cent.

Around half of R&D in Taiwan (a much higher proportion than in Korea) is financed by the government, though the contribution has come down over time. Private-sector R&D has been relatively weak because of the preponderance of small and medium enterprises (SMEs), which cannot afford the large minimum investments involved in much of industrial research. However, enterprise R&D has risen over time as some local firms have grown in size and become significant multinationals. Such R&D has been encouraged over the years by a variety of incentives: provision of funds for venture capital; financing facilities for enterprises that developed 'strategic' industrial products (of which 151 were selected in 1982 and 214 in 1987);[8] measures to encourage product development by private firms by providing matching interest-free loans and up to 25 per cent of grants for approved projects;[9] full tax deductibility for R&D expenses, with accelerated depreciation for research equipment; special incentives for enterprises based in the Hsinchu Science Park (with government financial institutions able to invest up to 49 per cent of the capital); and requiring larger firms (turnovers exceeding NT$300 m.) to invest (0.5–1.5 per cent of sales, depending on the activity) in R&D. The government also launched large-scale research consortia, funded jointly with industry, to develop critical industrial products such as a new generation automobile engine, 16M DRAM and 4M SRAM chips.

In sum, the main drive for rising R&D in Taiwan came, as in Korea, from the export orientation of the economy, combined with measures to reduce dependence on technology imports (below). However, Taiwan's 'lighter' industrial structure constrained the growth of private-sector R&D in comparison with Korea. In broad terms, both countries show the strong influence of strategic rather than static interventions on market failures.

IMPORTING FOREIGN TECHNOLOGY

Foreign technology is the primary input into industrial technology development in developing countries: facilitating access to it is thus a vital element of technology policy. However, the international technology market, while very active, is not like a market for physical products. In a normal market for physical products, sellers, prices and terms are relatively well known to buyers, the transaction is self-contained and can be completed at one go. Technology markets are different. They involve high search costs for buyers, especially new and small ones that are not engaged in international operations. The product may be difficult to define and value, and the market is often fragmented and oligopolistic, with information asymmetrically distributed between buyers and sellers. Once purchased, the effective use of the technology requires complementary physical and human inputs that the buyer may not have, and that may be difficult to locate and price. The sale may thus contain a number of different, complex elements, that take time to transfer and develop locally. Bargaining power may be the most important determinant of the price and terms of the sale, which may be hemmed in by restrictive conditions set by the seller.

These inherent imperfections in technology markets may call for policy intervention, and many developing countries have mounted such intervention. However, the scope for legitimate intervention goes beyond improving technology access and transfer in a general sense: the *mode of transfer* is also of concern. Technology imports can take several forms: purchase of equipment by itself (or along with know-how and technical assistance), contracts for blueprints and patents and other licensing arrangements, hiring of consultants, copying and 'reverse engineering' by the importing country, and so on. At one end, it can be entirely *internalised* by the seller of technology, as with wholly foreign-owned direct investments (in which the investor provides the equity capital, sets up and operates the technology and retains control over all subsequent expansion, upgrading and investments). At the other, it can be fully *externalised*, with the buyer putting together its own package of equipment, know-how and other inputs, with no long-term contractual arrangement with a foreign supplier (an extreme example would be a firm buying machinery on open markets and 'reverse engineering' a product design with no explicit technology agreement). In between can lie different equity or non-equity contractual relationships: joint ventures, licensing, franchising, subcontracting or turnkey contracts. Most countries, developed or developing, would have a mixture of these forms coexisting with each other, reflecting different vintages of technology, different levels of domestic technology development or enterprise strategies, different industrial structures and so on.[10]

Most industrialising countries in Asia have tried to improve their access to foreign technologies by helping their enterprises to obtain information on foreign technologies, and to improve upon the terms of the transfer. However, different countries have had very different strategic

Table 3.2   FDI inflows into Asian economies ($ million)

Country	1982–7 (average)	1988	1989	1990	1991	1992	1993
Hong Kong	1014	2627	1077	1728	538	1918	1667
China	1362	3194	3393	3487	4366	11156	27515
Korea	253	871	758	715	1116	550	N/A
Singapore	1605	3655	2773	5263	4395	5635	6830
Taiwan	306	959	1604	1330	1271	879	917
Indonesia	282	576	682	1093	1482	1774	2004
Malaysia	844	719	1668	2332	3998	4469	4351
Thailand	287	1105	1775	2444	2014	2116	1715

Source: UNCTAD, *World Investment Report, 1994.*

approaches to indigenous technology development and to internalised versus externalised imports. Table 3.2 shows inflows of foreign direct investment into the NIEs and 'new NIEs' and illustrates the varying extent of reliance on this mode of technology transfer.

Five broad types of strategies of technology import may be distinguished in the region. Some differ on their utilisation of FDI within a broadly open-door policy to MNCs, while others differ in their basic strategy on the entry of foreign investors.

Hong Kong

Hong Kong is exceptional in East Asia in that it always maintained a free-trade regime and did not explicitly influence technology imports. Its industrial and manufactured export growth was sparked off by an influx, after the communist take-over, of seasoned textile and other entrepreneurs and technicians from mainland China. This led to the emergence of dynamic small and medium-sized exporters specialising in relatively simple labour-intensive activities such as textiles, garments, toys and simple consumer electronics aimed at world markets. Given the initial endowment of skills and experience, they obtained the technologies needed in mainly externalised forms, primarily capital goods; the export drive had little need for advanced proprietary knowledge obtainable in internalised forms. The economy's colonial administration, its long experience of *entrepôt* trade, and the strong presence of expatriate-run trading, finance, property and other enterprises (the 'Hongs'), strengthened the initial base of manufacturing skills with an advanced (and possibly unique) physical, administrative, trading and financial infrastructure for export activity.

Despite open-door policies to FDI, Hong Kong's industrial development was dominated by indigenous firms (which accounted for over four-fifths of its manufactured exports). MNCs went mainly into service activities, while those that entered manufacturing specialised in slightly more advanced technologies within the same labour-intensive set of activities as local firms. The government made no effort to target high-technology FDI or to induce industrial deepening and technological upgrading. Technological information needs were thus relatively simple, and were fulfilled by scouting international suppliers of equipment (greatly helped by the liberal trading environment and the Hongs), by growing contacts with export markets, and by some government technology-support institutions (on which, more below). The presence of foreign buyers was always a vital source of technological information and assistance. Over time there was significant upgrading of equipment and products within the low-technology activities that the colony started with, but there was relatively little entry into complex and research-intensive technologies that the other NIEs were targeting.

In this context, the government did help technology imports *indirectly*. Its export-promotion measures enabled local firms to contact foreign buyers through the Hong Kong Trade Development Council, a well funded and staffed body which provided information on overseas markets, and assisted foreign buyers to establish contacts with local suppliers. In addition, technology imports and diffusion were undertaken by the Hong Kong Productivity Council (below), and a textile-design and training institute helped to upgrade production and design skills for the main export activity. However, the colony's relatively low wage increases (Hong Kong had the lowest rate of wage increase of the four Tigers over time) led to a continuing specialisation in relatively simple labour-intensive products. However, even this rate of increase, combined with rising land costs, and th2e lack of industrial deepening policies led to massive deindustrialisation in Hong Kong. Thus, employment in manufacturing declined by 50 per cent and the share of manufacturing in GDP from 25 to 7 per cent over 1980–95, and the process continues.[11]

The Hong Kong strategy on technology imports suggests that, where the appropriate skills, institutions, and infrastructure are present, free-market policies can stimulate investment and competitiveness in activities with 'easy' learning. However, where learning costs are high and specialised skills and information are involved (as in the industrial and technological deepening), more selective policies are required after the initial, 'entry-level' stage of industrial development. Hong Kong now lags behind other Asian NIEs in high-technology areas of electronics. Partly as a result of its hands-off industrial strategy, Hong Kong has the lowest rate of manufacturing growth (around 2 per cent in the 1990s) among the NIEs, and its manufactured exports (excluding re-exports) have been declining rapidly while rising strongly in the others (see Chapter 1). The economy has continued growing by moving into services and relocating manufacturing facilities to lower-cost areas, but as far as industrial technological deepening and diversification are concerned its experience has not been impressive.

Singapore

Singapore has a much smaller economy than Hong Kong's, but has deepened its industrial structure and maintained industrial and manufactured export growth. It started with a strategic location and established *entrepôt* facilities, like Hong Kong, though with a smaller base of trading and financial activity. Though it had a tradition of shipbuilding and associated skills, Singapore had a weak entrepreneurial base and did not benefit from an influx of experienced Shanghai businessmen and technologists. Nor did it have access to a large, less-developed but culturally similar, hinterland to which it could sell its services. After a spell of import substitution, it switched to free trade and pursued growth through aggressively seeking and targeting foreign direct investment, while raising domestic resource mobilisation by various measures. Moreover, it chose to deepen its industrial and export structure, and used a number of selective interventions to move from labour-intensive to capital-, skill- and technology-intensive activities. Its technology-acquisition policy was directed at consciously acquiring, and subsequently upgrading, the most modern technologies in highly internalised forms. This allowed it to specialise in particular stages of production within global systems of MNC production, drawing on the flow of innovation generated by the firms and investing relatively little in its own innovative effort.

To attract foreign investment while inducing it to upgrade, Singapore invested heavily in education and training and in physical infrastructure. It developed an efficient, industrially-targeted, higher technical education structure,[12] together with one of the best systems in the world for specialised worker training. Its policies for attracting FDI were based on liberal entry and ownership conditions, easy access to expatriate skills, and generous incentives for the activities that it was seeking to promote. It set up the Economic Development Board (EDB) in 1961 to coordinate policy, offer incentives to guide foreign investors into targeted activities, acquire and create industrial estates to attract multinational corporations, and generally to mastermind industrial policy (Box 3.1). At times it deliberately raised wages to accelerate technological upgrading, though in the mid-1980s a sharp rise in wages was modified to restore competitiveness. The public sector has always played an important role in launching and promoting activities chosen by the government, acting as a catalyst to private investment or entering areas that were too risky for the private sector. While the main thrust of Singapore's technology-import policies has been to target FDI, in recent years the government has also sought to increase linkages with local enterprises by promoting subcontracting and improving extension services (below).

Box 3.1 Singapore's FDI strategy

The decision of MNCs about what new technologies to bring into Singapore were strongly influenced by the incentive system and direction offered by the Singapore government, responding (or anticipating through proactive planning and consultation) by providing the necessary skilled manpower in consultation with the MNCs. In many instances, it was the *speed and flexibility* of government response that gave Singapore the competitive edge compared with other competing host countries. In particular, the boom in investment in offshore production by MNCs in the electronics industry in the 1970s and the early 1980s created a major opportunity by ensuring that all the enabling supporting industries, transport and communication infrastructure, as well as the relevant skill-development programmes, were available to attract these industries to Singapore. This concentration of resources helped Singapore to achieve significant *agglomeration economies* and hence first-mover advantages got many electronics-related industries. An example is the disk-drive industry, where all the major US disk-drive makers have located their assembly plants in Singapore. These industries demanded not only electronics components and PCB-assembly support, but also various precision engineering-related supporting industries such as tool and die, plastic injection moulding, electroplating and others. These supporting industries were actively promoted by the government as part of a 'clustering' approach to ensure the competitiveness of the downstream industries. As labour and land costs rose, the Singapore government used the opportunity to encourage MNCs to reconfigure their operations on a regional basis, making Singapore their regional administrative headquarters and/or regional marketing/distribution/service/R&D centres to support manufacturing and sales operation in the ASEAN and Asia–Pacific region. To promote such reconfiguration, new incentives such as the regional headquarters scheme, international procurement-office scheme, international logistics-centre scheme, and the approved trader scheme were introduced.

The success of Singapore's strategy is reflected in that, despite its small size, it remains (in dollar terms) the second largest destination for FDI in Asia, after China. Its industrial and manufactured export growth continues at near double-digit rates. As measured by the proportion of earnings accounted for by products of high R&D intensity, Singapore's export structure is now the most 'high-tech' in the world, more so than Japan's. This does not mean that the R&D is being done locally (it is not), but that the Singaporean industrial sector is deploying very advanced technologies in making very advanced products. Again, the contrast to Hong Kong is noteworthy; the difference is traceable to their different technology strategies.

Malaysia

The Singapore strategy is being copied by some of the 'new NIEs', especially Malaysia. Of the South-East Asian countries, Malaysia has been the most successful in attracting high-technology MNCs into manufacturing, and its growth, like Singapore's, has been driven by internalised technology imports by an export-oriented electronics industry.[13] Malaysia's initial success in attracting electronics MNCs was based on targeting firms that were relocating the more labour-intensive assembly activities from Singapore. Malaysia offered good infrastructure, low wages, literate and trainable labour, a stable macroeconomy with a convertible currency, combined with efficient administration and generous incentives for export-oriented activity. It was able to pre-empt other developing countries in the boom of electronics-assembly activity in the early 1970s. This gave it considerable first-mover advantages. Once established in Malaysia, MNCs responded to rising wages by increasing automation and process upgrading rather than by 'footloose' behaviour. Malaysia is now adopting Singapore-style policies to induce MNCs to upgrade technologically by undertaking local design and development, and to strike deeper local supply linkages. Its investment-promotion agency is honing its incentive system to this end, though its lack of high-level technical manpower holds back technological upgrading to levels reached by Singapore.

Apart from its attraction of high-technology export-oriented MNCs, Malaysian technology-import policies differed from Singapore's. Malaysia has retained a protected import-substituting regime, and in the 1980s intensified it with a heavy-industry programme led by the public sector. It imposed various restraints on MNC entry into the domestic-oriented sector of industry, and within the local sector sought to promote Malay entrepreneurship and participation (the export-oriented sector was insulated from these policies). In the late 1980s some of the restrictions on foreign entry were relaxed after a recession in exports and fall in FDI, resulting in a resumption of vigorous growth in both. The import-substituting sector did not have many technology-import interventions, and little help was given to local enterprises in searching for and buying foreign technologies.

Taiwan[14]

Taiwan started on import-substituting industrialisation in the 1950s with a strong base of human capital and a large population of SMEs. As with Korea, it switched to export orientation in the 1960s, but retained protection and targeting to promote and guide industrial growth. It combined these with interventions in technology transfer to support technology development by local enterprises. It drew upon the whole gamut of technology imports, but changed the balance and the policy regime over time. In the 1950s, it sought to attract FDI within a liberal regime, with no discrimination by origin, destination (only services were restricted for foreign entry) or degree of ownership. In the 1960s, FDI was sought in labour-intensive industries like textiles, garments and electronics assembly. In the 1970s, with rising wages and a need to upgrade industry, the government targeted higher technology, discouraging labour-intensive FDI and favouring investments in automation, informatics and precision instruments. This targeting was strengthened in the 1980s, as high-technology industries were granted five-year tax holidays, accelerated depreciation of equipment, low tax rates for selected activities and duty-free imports of R&D materials and equipment.

Thus, as the industrial sector developed and technologies deepened over time, FDI policy in Taiwan became more discriminatory. The government exercised more detailed surveillance (often on a case-by-case basis) to ensure that the technology was in line with changing national priorities. It targeted emerging technologies, and placed strict conditions on investors to benefit the technology development of domestic firms. Where domestic firms were strong, FDI was actively discouraged; where they were weak, foreign firms were made to diffuse technology and contribute to local capabilities. With yet more development of local firms and capabilities, selectivity on FDI was relaxed but the guidance and support of technology development continued. In the meantime, Taiwanese firms themselves became major investors overseas, spurred by the need to relocate labour-intensive activities and an enormous balance-of-payments surplus.

Box 3.2   Singer and local-content provisions in Taiwan

When the Singer Sewing Machine Company started operations in Taiwan in 1964, there were several small sewing-machine manufacturers in the country, with poor technology and no standardisation, unable to compete in world markets. The government stipulated that Singer procure 83 per cent of parts and components locally within 1 year, provide local suppliers with standardised blueprints, provide technical experts to improve productivity, prepare materials specifications and inspect final products. Singer was to provide local sewing-machine producers with its own locally made parts at no more than 15 per cent above the price of parts imported from Singer's foreign suppliers. The company was also required to raise exports rapidly.

The company fulfilled all the requirements, sending several technical and management experts to Taiwan to train and upgrade local suppliers and organise the entire production system. It provided a wide range of technical assistance to competing local sewing-machine manufacturers free of charge. Suppliers were given standardised blueprints, enabling them to work to common specifications; they were also given measuring instruments and access to Singer's tool room and technical advice. Classes were conducted for parts suppliers in technical and management problems.

The result of the forced local-content policy was a significant transfer of technology, increase in backward linkages and upgrading of competitive capabilities for the industry as a whole. Within three years Singer was using only local parts (except for some needles), and by 1986 was exporting 86 per cent of its total output. Other local firms also became major exporters, as local parts became standardised and improved in quality. One reason for this striking success was that relatively little capital investment was entailed. The existing base of technological capabilities in the local suppliers made the transfer and upgrading of technology relatively rapid low-cost. This pattern was repeated in several industries over the years.

Source: Dahlman and Sananikone (1990).

The government sought to maximise benefits from FDI for local firms by promoting local sourcing and subcontracting – an exceptionally successful strategy for enhancing technological and skill linkages with foreign firms (Box 3.2 describes the case of Singer). This promotion was done by local-content rules, backed by provisions that foreign firms transfer skills and technology to subcontractors and raise the technological capabilities of local firms.

The Taiwanese government also played a *direct* role in developing advanced technologies, where it found that the private sector was unable to develop the necessary capabilities. Take semiconductors. By the 1970s, Taiwanese industry had fallen behind technologically in this industry, which provided a crucial input into its burgeoning electronics export sector. The lag arose mainly because local firms were too small to set up the capital-intensive facilities involved and to invest in developing the necessary skills. In 1976, the Electronic Research and Service Organisation (ERSO), part of the government's Industrial Technology Research Institute (ITRI), imported and started to develop process technologies for very large integrated circuits (VLSI). By mastering this technology and creating a base of technical skills, ITRI was able to spin off the first integrated-circuit manufacturer in Taiwan in 1982.[15] This firm (UMC) was able to conclude agreements with three Chinese-owned start-ups in Silicon Valley in the USA to develop advanced chip designs. This was successful, and UMC went public in 1985. In 1987, using VLSI technology from UMC, the government set up a joint venture (the Taiwan Semiconductor Manufacturing Company, TSMC) with Philips of the Netherlands and local private interests for wafer fabrication. TSMC grew rapidly, and supported the development of design and manufacturing capabilities in numerous small electronics firms. This further encouraged the entry of private companies in the production of semiconductors, microprocessors and related electronics products: the government had been able to catalyse technological development by its critical intervention.

Foreign firms accounted for a relatively small part of Taiwan's industrial and export success. Local enterprises, led by SMEs, led the export drive, first by using the 'Chinese connection' in Asia and then, as their horizons widened, by tapping Japanese trading companies and American mass-market buyers. In the 1960s, about 60 per cent of textile exports were sold through Japanese trading houses (the *sogo shosha*), and even today these handle between a third and half of Taiwanese exports; such are the economies of scale and information collection in world markets that small firms find it difficult and costly to export alone even after years of experience (this is in contrast to the Korean case, reviewed below, where the government sought to internalise these functions within local trading houses, part of the giant local conglomerates). US buyers grew more important over time, with the government facilitating contacts with small suppliers, with aggressive assistance from industry associations and other private organisations. In addition, there also emerged many (relatively small) local trading houses, which proved to be valuable sources of technical, design and marketing information to exporters. Large multinational producers, that sourced complex electronic and related products under OEM (original equipment manufacture, where the product is sold under the brand name of the buying company) arrangements in Taiwan, were even more significant sources of technology transfer.

The China External Trade Development Council (CETRA), set up by the government in 1970 and funded by a (0.6 per cent) levy on exports, was an important agent of export promotion. CETRA developed sophisticated computerised data banks on foreign markets, buyers and suppliers, providing a one-stop source of information on supply potential in Taiwan. Its Industrial Design and Promotion Department helped exporters to develop designs and packaging appropriate to different foreign markets. By 1989 the organisation had 700 staff and operated 28 branch offices overseas. Ethnic Chinese from Taiwan living in the USA were also a significant source of technology, investment and skill transfers. In recent years, investment by Taiwanese companies in the developed world has become a growing source of technology, along with strategic alliances with technology leaders (Hobday, 1995).

Korea

In accessing foreign technology, Korea preferred externalised technology imports even more strongly. It relied primarily on capital-goods imports, technology licensing and other technology-transfer agreements to acquire technology. FDI was permitted only when it was the sole way of obtaining the technology or gaining access to world markets.[16] Even then the government encouraged majority Korean-owned or equal joint ventures; in some cases foreign investors were even forced to sell out after the technology had been absorbed locally. As a result, Korea had the lowest level of reliance on FDI of almost any developing countries with a non-communist economy, with the exception of India (see below). In recent years, under pressure from its trading partners, the Korean government liberalised its foreign-investment law. Nevertheless, some restrictions exist, and the entrenched position of its *chaebol* now means that Korea remains a small recipient of FDI in industry. Its *chaebol* have become major MNCs in their own right, by far the most powerful and aggressive of Third World MNCs.

The government intervened often in technology imports to lower prices and strengthen the position of local buyers, but in a flexible way that did not constrain access to expensive know-how. The licensing policy was liberalised over the 1980s as the need for more advanced technologies increased. The regime encouraged reverse engineering and R&D by technology-importing firms to develop indigenous technological capabilities; many of the larger firms were later able to enter into collaborative ventures with world technology leaders on a more equal basis. In the field of plant and process engineering, the government stipulated that foreign contractors transfer their design knowledge to local firms, which quickly absorbed design technologies in some process industries.[17] Even more so than Taiwan, Korea was able to use imported technology to develop its domestic base of capabilities in advanced activities, rather than remaining passively dependent on inflows of foreign skills and innovations.

The *chaebol* soon developed sufficient international presence to manage their technology imports. However, SMEs had to be given continued assistance to search for and buy technologies overseas. As with Taiwan (and Japan), Korea compiled a data base on sources and prices of technology supply. This was linked to similar data bases overseas and provided on-line in major industrial centres. There was also a programme to increase SMEs' technological linkages with large firms (see below) but, unlike Taiwan, this was directed mainly at local large firms rather than at MNCs. As with the other export-oriented countries, foreign buyers were a valuable source of technology.[18] The government's export-promotion efforts contributed greatly to this mode of technology acquisition. Several promotion measures were involved, including financial incentives, export targeting, other pressures to export (such as access to import licences) and information support.[19]

The Korean Overseas Trade Agency (KOTRA) played a significant role in providing contacts and market intelligence, and bringing together foreign buyers and Korean suppliers. The *chaebol* themselves were instrumental in promoting exports by other firms via their trading arms, modelled on the Japanese *sogo shosha*. These had the financial and marketing strength to be able to substitute for foreign trading companies that small Taiwanese exporters had to rely on (above), and contributed to the superior ability of Korea to establish its own brand names in international markets.

India

Indian technology-import policies had many features in common with those of Korea, but with very different results. Until fairly recently, India restricted FDI, protected domestic industry, encouraged externalised technology imports and tried to raise indigenous innovative capabilities. However, unlike Korea, this was done in an environment of stifling inward orientation, detailed bureaucratic licensing of industry, limitations on the growth of large firms and groups, strong preferences for public enterprises, restrictions on imports of licensed technology, pervasive protection for domestic capital goods, few contacts with export buyers, and policies that made 'exit' extremely difficult. The licensing system led to widespread rent-seeking, the setting up of suboptimal-sized plants, and over-diversification of business houses.

The Indian regime also imposed severe constrictions on industrial access to new foreign technology.[20] Until the late 1980s FDI inflows into India were minuscule; unfortunately, other access to technology was also limited. The trade strategy meant that the role of foreign buyers was very limited; and the lack of OEM purchases meant that the engineering sector remained deprived of modern designs and know-how. Capital-goods production was itself highly protected until the late 1980s, and importers had to establish that there were no domestic suppliers available, largely regardless of price and technological features. This deprived industry of new technologies embodied in modern equipment and the supporting information that came with equipment supply (by contrast, Korea was the largest single importer of capital goods in the developing world in the 1980s, combining this with selective protection for domestic equipment producers). Technology licensing was subject to a detailed, complex and rigid approval process, with upper limits on royalty rates and duration of agreements. The importer had to establish the novelty of the technology licensed. Restrictive clauses attached to technology agreements were prohibited or strongly discouraged, lowering the quality of the technology imported. Local-content provisions were rigorous, and strongly enforced.

All these interventions in technology purchases meant that the extent and depth of technology inflows to Indian industry suffered. In combination with the trade-incentive regime, this led to large areas of uncompetitiveness and technological sloth in industry. This is not to say that no technological development took place; there were some 'good' firms in almost every industry, and there was active learning associated with operating new technologies in the difficult Indian environment.[21] Yet the extent of learning and dynamism was clearly far less than in Korea. The critical difference was that Korea had designed and implemented its policies very differently. Its aggressive export-oriented regime combined with detailed targeting and credit allocation had supported large private groups, and pushed them into acquiring the latest and best available technologies, setting up world-class plants, and upgrading them constantly to enlarge world market share. India's strategy was not as coherent and systematic; it certainly failed to provide the incentive needed for technology development.

FINANCING TECHNOLOGICAL ACTIVITY

The ability to finance investments in technology development by existing firms and new technology-based ventures is a crucial element of ITD. At low levels of industrialisation, when firms are small and using easy technologies with low capital requirements and limited possibilities of improvements, the absence of specialised technology finance may not be a major handicap. Working capital may cover most technology-development activities related to production engineering, quality improvement and productivity improvement. Even at this level, however, there is a risk that such financing will not meet needs for training and other activities not strictly classifiable as working-capital needs; in any case, small firms may not be able to offer commercial banks collateral even for working capital.

Box 3.3 Japanese financial support for SME technology

In Japan's combined public–private support system for SMEs, the government role is pervasive. Spending and loans by the national government for help to all small business (including non-manufacturing) amounted to about 4.4 trillion yen in 1989, or $31.2 billion at 140 yen to the dollar ($44 billion at current rates). Of this, only $1.4 billion appeared in the regular general account budget, which is supported directly by taxes. The rest was in the Fiscal Investment and Loan Programme, a capital budget often called the 'second budget' which derives its revenues from government trust funds and the country's huge government-subsidised postal savings programme. Altogether, spending for small business amounted in 1989 to nearly 5 per cent of the total regular and capital budgets of the national government. This sum does not include spending by prefectures, cities and city wards, which also contribute handsomely to programmes for small business, matching the national government's contribution in many cases. The Japanese national programmes for SMEs include both financial and technical assistance. In the 1980s, special attention was given to programmes aimed to help small business adopt high-tech equipment such as computerised machinery and robots.

Among the many services the government offers SMEs are a big programme of direct loans for operating funds or plant and equipment investment and a still bigger programme of government-guaranteed loans. In 1987, loans to SMEs via the three main government financing institutions amounted to 3.8 trillion yen, or $27 billion ($38 billion at current rates). The loan-guarantee programmes are larger – 52 nation-wide credit guarantee associations underwrote 7.8 trillion yen in loans to SMEs in 1987. The Japanese government provides about 20 times more financial aid to small business than the US government does.

Government loans are an essential source of financing for small start-up companies with no track record. The recent trend is for private loans to overtake government financing; government loans dropped from 13 per cent of all loans to SMEs outstanding in 1989 to 9 per cent in 1988. These figures understate the government's role in financing SMEs, since they omit loan guarantees. In addition, the Japanese government offers other SME 'measures' funded at about 225 billion yen ($2.3 billion at current rates) in 1987. One of these, the Equipment Modernisation Loan System, made 6000 loans in 1987 for 41 billion yen. The Equipment Leasing System provides equipment to very small firms (20 or fewer employees) on lease or on instalment purchase. The government is also a partner in quasi-private leasing companies that serve large and small companies.

Source: OTA (1990), pp. 161–3.

As industrial development proceeds, the financing gap grows more serious. ITD increasingly takes the form of long-term and risky investments in new technologies by existing firms and by new start-ups that do not have a track record. The financial system is generally not willing to finance such investments, except by very large firms that can cross-subsidise their R&D activity. Capital-market failures are recognised in developing countries, especially as far as the financing of smaller enterprises is concerned. Such failures are also accepted as being widespread in developing countries,[22] and they apply with greater force to technology financing. All the NIEs have adopted measures to provide technology financing and meet the special needs of SMEs.

First, however, it may be interesting to consider how Japan, the developed country with the largest and most dynamic set of SMEs, supports technology upgrading in the small-scale sector. Box 3.3 describes government SME programmes, taking the evidence from the US government's OTA.[23]

Korea

Korea's policies to selectively encourage activities and firms via credit allocation and subsidisation were inherent to its industrial policy from the start (Amsden, 1989; World Bank, 1993). As the industrial sector matured and entered more demanding areas of technology, and the government reduced the direct allocation of credit, its role in technology financing increased rather than decreased.[24] This emphasis was also aided by the fact that the emerging 'rules of the game' made other forms of subsidies and grants to industry unacceptable, while technology financing remained a permissible form of intervention.

The Korean government provided technology financing in the form of both grants and loans (often directed and subsidised). A variety of institutions, like venture-capital companies, banks, credit guarantee companies and others were used to channel funds to a variety of users in a variety of forms. These three forms of technology financing, subsidies, loans and institutional support, are described in turn.

Subsidies

There are three main forms of subsidies for technological effort: the Designated R&D Programme (launched in 1982), the Industrial Technology Development Programme (1987), and the Highly Advanced National Project (HAN) (1992). Together these have contributed large sums of money for research approved or targeted by the government, conducted by firms on their own, by research institutes on their own, and by firms in collaboration with research institutes (Box 3.4).

Box 3.4   Korean government's subsidies for technology development

- The Designated R&D Programme has, since 1982, supported private firms undertaking research in core strategic technology-development projects in the industrial area approved by the Ministry of Science and Technology. It funded up to 50 per cent of R&D costs of large firms and up to 80 per cent for SMEs. Between 1982 and 1993, this programme funded 2412 projects, which employed around 25,000 researchers at a total cost of around $2 billion, of which the government contributed 58 per cent. It resulted in 1384 patent applications, 675 commercialised products and $33 million of direct exports of know-how. Its indirect contribution in terms of training researchers and enhancing enterprise research capabilities was much larger. The value of grants under the programme in 1994 was $186 million, of which 42 per cent was directed at high-technology products like new speciality chemicals.

- The Industrial Technology Development Programme, started in 1987 to subsidise up to two-thirds of the R&D costs of joint projects of national interest (National Research Projects)[a] between private firms and research institutes. Between 1987 and 1993 this programme sponsored 1426 projects at the cost of $1.1 billion, of which the subsidy element from the government was 41 per cent. In 1994, the programme gave grants of $180 million (with 31 per cent going to high-technology products), a significant increase from $69 million in 1990.

- The Highly Advanced National Project (HAN) was launched in 1992 to support two activities: the development of specific high-technology products in which Korea could become competitive with advanced industrial countries in a decade or two (Product Technology Development Project), and the development of 'core' technologies considered essential for the economy in which Korea wanted to achieve an independent innovative base (Fundamental Technology Development Project). So far 11 HAN projects have been selected, and during 1992–4 the government provided $350 million of subsidies for them. In this brief period, the programme resulted in 1634 patent applications and 298 registrations.

[a] These National Projects are mentioned below in the section on the technology infrastructure.

Source: Song (1995).

Loans

The Korean government set up three funds to provide loans, usually at subsidised rates,[25] for technology development. The *first* was the Industrial Development Fund, providing low-interest loans for long-term productivity improvement and technology upgrading in high-technology industries. Several banks were used to channel the funds, which could go up to 70 per cent of the approved projects for large companies and up to 100 per cent for SMEs. The loans were given for five years, with a two-year grace period, and an interest rate of 6.5 per cent. The total funds disbursed over 1990–94 came to around $618 million. The *second* was the Science and Technology Promotion Fund, started in 1993 to fund firms and research institutes undertaking HAN projects (noted above). Loans can go up to 80 per cent of the total value of the project, up to $1.3 million per project and $3.8 million per firm. The loans are for seven years, with a grace period of three years and an interest rate of 6 per cent. In its two years of operation the fund has offered $255 million. *Thirdly*, an SME Foundation Formation Fund was set up as recently as 1994 to support technology development and environmental investment by smaller firms. The fund could finance 100 per cent of approved projects at an interest rate of 8.5 per cent over ten years, with a grace period of three years. In 1994 this fund offered $400 million.

Financial institutions' technology financing

Korea has the largest and most successful *venture capital* industry in the developing world. Starting with the launching of the Korea Technology

Development Corporation (KTDC), a joint effort by the government and the *chaebol*, in the early 1980s, several private venture-capital funds were set up. There are 58 venture-capital companies in Korea today, which disbursed loans and investment funds amounting to $3.5 billion during 1990–94 (85 per cent of this was in the form of loans).

A number of *banks* (Korea Development Bank, Industrial Bank of Korea, the Kookmin Bank, the Korea Long-Term Credit Bank and others) lend money to firms and research institutes for technology development. The state-owned KDB, for instance, offers three kinds of finance: Technology Development Loan, High-Technology Industry Promotion Loan and Production Technology Development Loan. These three instruments lent $3.4 billion during 1990–94, with 40 per cent going into the High-Technology Industry Promotion programme. Both this programme and the Production Technology Development Loan are made to firms approved by the Ministry of Trade, Industry and Energy; finance is provided for eight years with a three-year grace period and a subsidised interest rate of 8 per cent. The Industrial Development Bank of Korea (IDB) offers Technology Development Loans for SMEs, which amounted to $560 million during 1990–94. These loans are for developing new technologies or improving upon imported technologies, and IDB offers up to 100 per cent of the cost of the project at 8.5 per cent interest (over ten years with a three-year grace period). Other banks also offer similar loans to SMEs.

The Korea Technology Credit Guarantee Fund (KTCGF) offers *credit guarantees* for loans made to help firms develop or commercialise new technology. It concentrates on SMEs (firms with under 1000 employees) in new technology industries, as well as research institutes that need funds for technology development. The total value of its guarantees between 1990 and 1994 came to about $8 billion. The fee charged is 1 per cent of the value guaranteed for SMEs and 1.5 per cent for larger companies.

The scale of technology financing in Korea is truly impressive, though the government feels that it is still inadequate for its needs. This accounts for the constant setting up of new schemes, targeted at smaller firms, and the fostering of collaboration with research institutes. The figures also indicate that there is tremendous technological dynamism in the SME sector, though the *chaebol* continue to account for the bulk of R&D expenditures. The extent of selectivity in technological activity remains very high, with no remission in the strategy of identifying and targeting specific areas for research activity.

Taiwan

Taiwan also developed a comprehensive system for financing technology activity. In the early 1980s, the government found that the financial system was failing to meet the need of technology-based enterprises. It set up a capital investment fund of NT$800 million in 1983, which it augmented in 1991 by a second fund of NT$1.6 billion. By mid-1993 it had 23 venture-capital companies, which had invested some NT$9 billion (US$340 million) in nearly 400 companies in high-technology industries (nearly half the funds went into two activities, information and electronics).

It is also of interest here to note an example of Taiwan's financial support of industrial restructuring, in the *technological upgrading of the textile industry*. Textiles have long been the second largest export for Taiwan (reaching US$12 billion in 1993). Though labour-intensive garments production has been largely relocated to lower-wage countries, and the industry within Taiwan is dominated by synthetic fibres, there remain large sections of the industry that need to be upgraded. Despite their international exposure and large export earnings, the firms concerned have been unable to upgrade sufficiently under free markets. As a result, the Taiwanese government embarked in the late 1980s on a major financing programme for restructuring the industry. The objective was to raise exports to $20 billion over the 1990s, help producers to raise the quality of their products and design, and move the more labour-intensive processes overseas.

The Industrial Development Bureau of the Ministry of Economic Affairs developed a $95.4 million programme, of which 95 per cent was to be provided as grants to private firms to speed up technological renovation, encourage R&D, and improve design capabilities and other skills. Over 250 textile plants were to receive financial and technical assistance under this programme, enabling them to import the latest automated equipment, train their staff in the new technologies and develop new design skills. A number of other agencies, both public and private, were to be involved in this restructuring exercise. The Taiwan Textile Federation and the CETRA Industrial Design Centre were to provide information, training and trade fairs. The China Productivity Centre was to provide technical teams to visit plants and advise on automation. Banks were to provide low-interest loans to SMEs to move their facilities overseas, with special credit lines (up to $60,000 each) to import new equipment. This example illustrates the strategic use of finance by the Taiwanese government to target and support, not a new high-technology activity (which it was also doing) but a mature industry which faced market failures in expensive restructuring.

Malaysia

The Malaysian case illustrates some *problems* inherent in launching technology finance in a newly industrialising economy. We note two aspects of finance – *venture capital* and the *official programmes* for encouraging technology development – in which the government has been active.

Venture capital

The venture-capital (VC) industry, owned privately but promoted by the government, is growing well. In 1994 its investments came to US$58 million, reaching 67 enterprises, an impressive increase from the US$25 million to 32 companies in the previous year.[26] Around 56 per cent of the 1994 total went into manufacturing, with two-thirds directed at electrical and electronics firms. While this suggests a desirable focus on high-technology activities, a look at its activities gives a different picture. The bulk of VC investments (65 per cent) was directed to the funding of *acquisitions* and *bridge financing*; only 26.7 per cent was invested in company start-ups. Of this, relatively little went into financing genuinely technology-based new ventures, and most was directed at expansion projects by large, credit-worthy firms. In addition, the average size of venture-capital investments was very small. The funds were clearly unwilling to take risks, and probably unable to assess the viability of technology-based projects. The finance was not really reaching technology-based activity. There was also probably a small demand for technology loans from new entrepreneurs, though it seems from interviews that existing demand in Malaysia was not being met by the supply.

Official schemes

The government-sponsored Malaysia Technology Development Corporation (MTDC) finances the commercialisation of technology produced by universities and research laboratories. MTDC's output has been modest, however, since the output of industrially useful technology from these institutions is fairly low.

There exist three schemes to promote technology upgrading by SMEs:

- Vendor Development Programmes (VDPs) provide potential supplier firms with indirect access to foreign technology via an input-purchasing 'anchor' company. The government plays a catalytic role in these schemes, but the anchor companies (mainly public-sector firms like Proton and some MNCs) have to carry the full costs of the training and related activities undertaken to upgrade SME production and absorptive capabilities. Partly as a result of the costs involved to the anchor firms, the impact of this scheme has been limited; however, Proton has succeeded in supporting the growth of a large SME supplier system.

- The Industrial Technical Assistance Fund (ITAF) provides matching grants for SMEs (of up to US$230,000 per firm) to undertake consultancy and feasibility studies, product development and design, quality and productivity improvement, and market development. The uptake of the fund has been disappointing, and by the end of 1994 (three years after the scheme started), grants came to US$7.4 million,[27] only 37 per cent of the total available. This weak response can be traced to several factors: SMEs are not fully aware of the benefits of the programme or even the need for upgrading; the procedures are cumbersome (the scheme involves several agencies) and demanding for firms with little spare manpower; SMEs find it difficult to formalise their technological needs as required by the scheme; and some enterprises are unwilling to expose themselves to authority.

- The Soft Loan Scheme for Modernization and Automation, to stimulate the purchase of new machinery and equipment by SMEs, has again been disappointing, for the same reasons noted above.

In principle, these schemes seem well-geared to meeting the technological needs of SMEs. Their limited impact on technological upgrading must therefore reflect weaknesses in their design and delivery and the lack of capability on the part of enterprises. The experience of Taiwan suggests (below) that technology assistance to SMEs needs to be much more aggressive in reaching out to enterprises and providing an integrated and customised package of technological, financial, management and marketing assistance. It also needs to have simple and quick procedures that minimise time and information requirements on the part of entrepreneurs. The Malaysian schemes have, by contrast, been rather passive and bureaucratic; and Malaysian SMEs lack the ability to utilise external assistance effectively.

Hong Kong

The deindustrialisation and low technological status of Hong Kong, referred to earlier, has recently provoked the launching of some schemes to finance and promote technological activity. The mainland Chinese government is also contemplating a more active technology policy after it takes the colony over in 1997. This reaction provides insights into the outcome of the *lack* of selective industrial and technology policies in an otherwise successful economy. Since detailed information on recent policies is lacking, this section relies on recent press reports on this subject. To quote,

> The government isn't watching passively as Hong Kong's manufacturing sector withers. Instead, the traditionally *laissez-faire* colony is trying both to shore up what is left of the domestic manufacturing base and to bolster the colony's position as the gatekeeper of Chinese manufacturing.... As part of the new interventionist tilt, the industry department on May 9 [1994] gave preliminary approval to spend HK$140 million (US$18 m.) on 39 projects designed to upgrade Hong Kong's industrial structure. These are the first projects approved under a scheme to spend some HK$200 m. annually to improve technology. Though far short of the degree of intervention in regional powerhouses Singapore, Taiwan and South Korea, it marks a notable turn-around for the colony.... A typical project aims to make palladium-coating technology available for the colony's watch and jewellery makers. The two year HK$4 m. project is being conducted by the Hong Kong Productivity Council in conjunction with the Hong Kong Watch Manufacturers Association. The arrangement – where projects are tied to trade associations or universities – is characteristic of the strategy to fund projects that would be too costly for individual firms to conduct, but essential to their survival ...
>
> The Hong Kong Industrial Technology Centre represents one of the colony's most tangible commitments to supporting technology-oriented companies. The centre is intended to promote technology activities in the colony by nurturing small companies and providing relatively inexpensive state-of-the-art facilities for more established ones.... The government package supporting the centre includes a one-time grant of HK$250 m., the ability to tap a HK$188 m. credit line (at 7% a year) and 5,700 square meters of land next to one of the colony's busiest train stations ...[28]

In order to encourage this trend [investment in new equipment], the government has toned down its *laissez faire* inclinations to permit a new applied research and development scheme. This is a $HK200m. fund, which will match the investment of any start-up company which fulfils certain criteria, in exchange for an equity stake. This represents the first step towards direct government funding for research and development, and by implication, the creation of a government industrial policy. . . . The government . . . has provided a $HK250 m. grant for an Industrial Technology Centre, a new University of Science and Technology was recently completed, and plans have been initiated for a $HK2.8 billion science park (the government's contributions would be much lower). In addition, its industrial estates in Tai Po and Yuen Long provide land for high-tech industries at significantly below market rates.[29]

The Chinese government is contemplating a more interventionist stance to R&D in Hong Kong. It wants to nurture high-technology R&D, both to enhance its competitive advantage and to act as a channel for more advanced technology transfer to the mainland.[30] Its policy makers intend to reverse its decline in manufacturing and its lack of an advanced technology base. Its enterprises are considering the advantages of Hong Kong as a base for developing and selling technology. For instance, China Aerospace International, a rocket manufacturer, plans to have (as yet unspecified) high-technology products developed in Hong Kong research centres. Chinese universities are setting up commercial subsidiaries that are seeking listing in Hong Kong to widen their international reach.

HUMAN RESOURCES FOR TECHNOLOGY DEVELOPMENT

The significance of human resources for technology development hardly needs emphasis. All Asian NIEs are acutely conscious of the need to create high-level technical, engineering and scientific skills to undertake technological activity and engage in R&D, and most governments attach the highest priority to education. The extent of technical skill creation is, however, highly variable between countries.

Table 3.3 illustrates some of the differences. The NIEs with the strongest technological ambitions, Korea, Taiwan and Singapore, have invested most in training scientists and engineers. Their governments have set up new universities and directed curricula towards technical subjects, encouraged foreign education and attracted back trained

Table 3.3   Educational enrolments for technology

| | Total | Tertiary Level Enrolment | | Vocational Training Enrolment |
| | | All S&T subjects | Engineering only | |
	(% of age group)		(% of population)	
H. Kong	20	0.50	0.25	0.79
Indonesia	10	0.08	0.06	0.17
Korea	40	0.96	0.58	1.93
Japan	31	0.43	0.37	1.17
Malaysia	7	0.15	0.07	0.17
Singapore	14(est.)	0.90	0.61	0.40 (1980)
Taiwan	37	0.92	0.68	2.12
Thailand	16	0.16	0.09	0.80

Sources: UNESCO, *Statistical Yearbook* (1994); World Bank, *World Development Report*; Taiwan, *Statistical Data Book 1994*.

nationals. While it is difficult to compare the quality of the education provided, some international tests of numeracy suggest that the NIEs and Japan have the best training at the school level in mathematics. Moreover, governments have not always provided substantial direct funding for higher education: in Korea, for instance, the bulk of higher education is funded privately, with the government playing a guiding and catalytic role. Let us take some specific policies.

Singapore

Singapore's system for meeting the skill needs of industrial employees, one of the best in the developing world, is worth describing at some length. Singapore is a regional leader in employee training programmes held outside the firm. The Vocational and Industrial Training Board (VITB) established an integrated training infrastructure which has trained and certified over 112,000 individuals, about 9 per cent of the existing workforce, since its inception in 1979. The VITB administers several programmes. The Full-Time Institutional Training Programme provides broad-based pre-employment skills training for school leavers. The Continuing Skills Training Programme comprises part-time skills courses and customised courses. Customised courses are also offered to workers, based on requests from companies, and are specifically tailored to their needs. Continuing Education provides part-time classes to help working adults.

VITB's Training and Industry Programme offers apprenticeships to school leavers and ex-national servicemen to undergo technical skills training while earning a wage. The programme consists of both on-the-job and off-the-job training. On-the-job training is carried out at the workplace where apprentices, working under the supervision of experienced and qualified personnel, acquire skills needed for the job. Off-the-job training includes theoretical lessons conducted at VITB training institutes or industry/company training centres. Unusually, the government collaborated with MNCs (one Indian, one German and one Dutch) to set up these centres jointly, funding a large part of employees' salaries while they are being trained (see below) in state-of-the-art complex manufacturing technologies. Later the Singapore government worked jointly with foreign governments (Japan, Germany and France) to provide technical training.[31]

Under the Industry-Based Training Programme, employers conduct skills training courses matched to their specific needs with VITB assistance. VITB also provides testing and certification of its trainees and apprentices as well as trade tests for public candidates. The Board, in collaboration with industry, certifies service skills in retailing, health care, and travel services. Using various grant schemes, the National Productivity Board's Skills Development Fund (SDF) created 405,621 training places in 1990. The initial impact of the programme was found mostly in large firms; however, efforts to make small firms aware of the training courses and provide support for industry associations has increased SDF's impact on smaller organisations. SDF is responsible for various financial assistance schemes to help SMEs finance their training needs and to upgrade their operations. It has also introduced a Development Consultancy Scheme to provide grants to SMEs for short-term consultancy for management, technical know-how, business development and manpower training.[32]

The Training Voucher Scheme supports employers with training fees. This scheme enabled the SDF to reach more than 3000 new companies in 1990, many of which had 50 or fewer employees. The Training Leave Scheme encourages companies to send their employees for training during office hours. It provides 100 per cent funding of the training costs for approved programmes, up to a maximum of $20 per participant hour. In 1990, over 5000 workers benefited from this scheme. The success of the Skills Development Fund is due in part to a strategy of incremental implementation. Initially, efforts focused on creating awareness among employers, with *ad hoc* reimbursement of courses. The policy was then refined to target in-plant training, and reimbursement increased to 90 per cent of costs as an additional incentive. Further modifications were made to encourage the development of corporate training programmes by paying grants in advance of expenses, thus reducing interest costs to firms.

Singapore's interventions in education went well beyond what may be regarded as meeting existing deficiencies in the system – they were an integral part of its strategy of industrial and technological deepening. They created an entirely new set of basic endowments to accommodate the activities and FDI that the government targeted, and they were selected very narrowly.

Korea

Korea has among the highest levels of educational attainments relevant to industry of any developing country, the only close competitor being Taiwan. Its secondary and tertiary-level enrolments (at 90 and 40 per cent respectively) are at developed-country levels. Dropout rates are very low and the quality of the education, as judged by international comparisons of numeracy and science tests, is very good at imparting numeracy. It has impressive levels of vocational training enrolments, and encourages significant in-firm training of employees. It has the highest relative enrolments in science and technology subjects at universities of the East Asian countries, a conscious strategy of the government to raise ITD to levels of developed countries. The government's manpower planning was based explicitly on projecting the need for high-level technical manpower with highly industrialised nations like Japan, Germany and the USA.

The education of high-level technical manpower was promoted by the setting up of institutions like KAIST (Korea Advanced Institute of Science and Technology) at the post-graduate level and KIT (Korea Institute of Technology) at the undergraduate level. These were aimed at exceptionally gifted students, while the normal university system catered to the normal run of science and engineering training. KAIST turned out a total of 6652 graduates between 1975 and 1990, of whom 832 were PhDs, and the rest MScs. An example of educational targeting in support of industrial strategy is the policy of upgrading electronics design skills. In 1988 the Korean government began funding the Seoul National University's semiconductor laboratory to train a new generation of chip designers. The laboratory mounted annual programmes for about 200 students and employees of private firms, and some of the work for Korea's highest-profile electronics project, the development of a 64 megabit DRAM chip, has been conducted at this laboratory. The laboratory receives about Won 600 million from the

Education Ministry. About 70 per cent of the money is distributed to students and faculty at smaller universities for use in related research efforts. The laboratory, with about Won 10 billion of equipment, is also involved in materials analysis.

Korea has also strongly encouraged in-firm training. The government levied a 5 per cent payroll tax on large firms, refundable if they undertook employee training in approved programmes. While such payroll taxes are found in many countries, the level set in Korea was exceptionally high (most others range around 1 per cent). This reflected the sharply rising needs for new skills with the Korean push into heavy and high-technology industry. It probably also reflected the initial reluctance of firms to invest in employee training, since there was no tradition of in-firm training, and the lack of a lifetime employment system (unlike Japan, where firms invest heavily in upgrading employee skills) reduced employer's ability to appropriate returns to these investments.

Malaysia

Malaysia suffers severe shortages of high-level technical manpower, seriously constraining its upgrading into higher-technology activities. However, it provides an interesting example of industry-led training: the *Penang Skill Development Centre* (PSDC). Penang has a concentration of high-technology activities, with many major electronics MNCs engaged in export-oriented activities. The PSDC was launched in 1989 in response to the growing skill shortages. The initiative, land and some financial support came from the Penang State Development Corporation, which induced three leading US electronics MNCs to participate in the venture. The MNCs formed a steering committee, giving finance and full access to their own training programmes, materials and methods. Other MNCs and local firms then started to participate and private industry continued to play a leading role in the institution, developing a strong sense of 'ownership' in its activities. PSDC borrowed trainers from these companies, and devised a range of training programmes suited to their needs. The full cost was charged for its services, and the programmes were continually upgraded and adapted to evolving skill needs. The centre is nevertheless entirely autonomous in its operations and decision making.

The Malaysian government offered 200 per cent tax deduction for training costs to firms that sent their personnel to the PSDC (and to other approved training centres) for training; this scheme has been recently replaced by the training levy mentioned above for large firms. It also allowed the centre to bring in, without formal work permits, high-level foreign trainers (however, 40 per cent of PSDC's training is done by local staff, and the proportion is rising steadily). The equipment needed for training was often obtained at no cost, from equipment manufacturers hoping to increase sales of their products. Some firms in Penang moved their own training equipment to the PSDC, releasing space in their plants.

From employee training, PSDC has recently moved into pre-employment training, also paid for by companies seeking skilled workers. The centre has started to give scholarships to school leavers and is now producing its own teaching materials. The PSDC has 58 member companies, 80 per cent of them foreign. Its annual budget (1994) was around US$385,000, of which only $48,000 came from the state government. It has conducted around 6300 courses, attended by about 9500 participants. Salaries paid to trainers are on a par with the private sector, and they are hired on contract without tenure. The Board is dominated by private industry, with some representation by technology institutions but none from the central government. Industry continues to give grants and regards its contribution as extremely valuable. The Malaysian government is now emulating the PSDC model in every other state of the country, with some degree of success.

TECHNOLOGY INFRASTRUCTURE SUPPORT

The technology infrastructure includes *MSTQ* (metrology, standards, testing and quality) *institutions, public research institutions and university research* concerned with industry, and *technical support and extension* for SMEs. Many of their services are 'public goods', with large externalities and therefore difficult to price on market terms, yet they are essential for technology development and diffusion. For instance, public research institutes and universities undertake basic research that does not yield commercial results in the short term, but provides the long-term base of knowledge for commercial R&D. Quality, standards and metrological institutions provide the basic framework for firms to communicate on technology, and keep the basic measurement standards to which industry can refer. Extension services help overcome the informational, technical, equipment and other handicaps that SMEs tend to suffer. The provision of these services thus faces market failures of the static sort that every government, regardless of its political ideology, has to remedy.[33] However, strategic policies can also be involved, since infrastructure institutions can be used to spearhead entry into new areas of technology targeted by the government. This was the case with some of the NIEs.

The NIEs all invested heavily in the technology infrastructure. Each made concerted efforts, for instance, to promote the International Standards Organisation's ISO 9000 quality-management standards,[34] increasingly important in export activity. In late 1993, for instance, the numbers of ISO 9000 certified companies were as follows: Hong Kong 10, Indonesia 8, Malaysia 224, Singapore 523, Taiwan 96, Korea 87 and Thailand 9. By late 1995, the numbers were much larger: Malaysia around 700, Hong Kong around 200, Singapore over 1000, and Indonesia 60. The region is now the furthest advanced in the developing world in ISO accreditation. Its standards bodies are playing aggressive roles in propagating the advanced quality management, and providing assistance and consultancy services. Some examples of other infrastructure activities by the NIEs are as follows.

Hong Kong

Despite its *laissez faire* approach to industry, the colony provides strong technical support to its SMEs through the *Hong Kong Productivity Council* (HKPC).[35] HKPC was the first support institution of its kind in the region, started in 1967 to help the myriad small firms that constitute the bulk of the industrial sector. Its focus has been to help firms upgrade from declining labour-intensive manufacturing to more advanced, high value-added activities. It provides information on international standards and quality and gives training, consultancy and demonstration services on productivity and quality to small firms at subsidised rates, serving over 4000 firms each year. Its on-line information-retrieval system has access to over 600 international data bases on a comprehensive range of disciplines. Its technical library takes over 700 journals and has over 16,000 reference books.

The HKPC acts as a major technology import, diffusion and development agent for all the main industrial sectors in the economy. It first identifies relevant new technologies in the international market, then builds up its own expertise in those technologies, and finally introduces them to local firms. Successful examples of this approach include surface-mount technology and 3-D laser stereo-lithography. HKPC has also developed a number of CAD/CAM/CAE systems for the plastics and moulds industry, of which over 300 have been installed already.

HKPC provides a range of management and technology-related courses, reaching some 15,000 participants per annum. For firms unable to release staff, it organises in-house training programmes tailored to individual needs. To help the dissemination of information technology, the council has formed strategic alliances with major computer vendors, and provides specially designed software for local industry, consultancy and project management in computerisation. HKPC provides consultancy services in ISO 9000 systems, and has helped several firms in Hong Kong to obtain certification. It assists local firms in automation by designing and developing special-purpose equipment and advanced machines to improve process efficiency.

HKPC is a large organisation, with over 600 consultants and staff, a laboratory and a demonstration centre that can show the application of new technologies (in CAD/CAM, advanced manufacturing technology, surface-mount technology, micro-processor technology, rapid prototyping and so on). In 1993–4, it undertook 1354 consultancy and technology-assistance projects, trained over 15,000 people and undertook 2400 cases of manufacturing support services. Because small firms experience difficulties in getting information on, and adopting, new technologies, and are exceptionally averse to the risk and cost involved, the HKPC has always had to subsidise the cost of its services. Despite the growth in the share of revenue-earning work and its withdrawal from activities in which private consultants have appeared, the government still contributes about half its budget. It is important to note that technological information market failures and the need for subsidised services occur even in a highly export-oriented economy, with highly developed financial services, like Hong Kong.

The Hong Kong government also supported local design capabilities by joining the private sector in starting a school of design. It financed the Hong Kong Design Innovation Company from the government because private-sector design services were lacking and local firms were not aware of their value. Over the four years of its existence (mainly on government financing) this value has been recognised, but the HKDIC (now under the HKPC) is still not financially self-supporting. Nevertheless, the growth of garment-design capabilities in Hong Kong has helped its exporters to upgrade their products and start to establish their own brands in international markets.

Singapore

Two aspects of Singapore's technology infrastructure programmes are worth noting. The first is its policies to help SMEs, illustrating the more static aspects of technology policy. The second is its use of public research institutes to catalyse private R&D in selected areas – these are strategic policies.

SMEs

In 1962 the EDB launched a programme to help SMEs modernise their equipment with funds provided by the UNDP. In the mid-1970s several other schemes for financial assistance were added; of these, the most significant was the Small Industries Finance Scheme to encourage technological upgrading (Soon, 1994). The 1985 recession induced the government to launch stronger measures, and the Venture Capital Fund was set up to help SMEs acquire capital through low-interest loans and equity. A Small Enterprises Bureau was established in 1986 to act as a one-stop consultancy agency; this helped SMEs with management and training, finance and grants, and coordinating assistance from other agencies. In 1987, a US$519 million scheme was launched to cover eight programmes to help SMEs, including product-development assistance, technical assistance to import foreign consultancy, venture capital to help technology start-ups, robot leasing, training, and technology tie-ups with foreign companies.

In addition, the Singapore Institute of Standards and Industrial Research (SISIR) disseminated technology to SMEs, and helped their exports by providing information on foreign technical requirements and how to meet them. The National Productivity Board provided management advice and consultancy to SMEs. The Technology Development Centre (TDC) helped local firms to identify their technology requirements and purchase technologies; it also designed technology-upgrading strategies. Since its foundation in 1989, the TDC has provided over 130 firms with various forms of technical assistance. It also administers the Small Industry Technical Assistance Scheme (SITAS) and Product Development Assistance Scheme to help firms develop their design and development capabilities. It gave grants of over $1 million for 29 SITAS in the past 5 years, mainly to local enterprises. Its earnings have risen to a level where its cost-recoverable activities are self-financing.

The EDB encouraged subcontracting to local firms through its Local Industries Upgrading Programme (LIUP), under which MNCs were encouraged to source components locally by 'adopting' particular SMEs as subcontractors. In return for a commitment by the MNCs to provide on-the-job training and technical assistance to subcontractors, the government provided a package of assistance to the latter, including cost-sharing grants and loans for the purchase of equipment or consultancy and the provision of training. By the end of 1990, 27 MNCs and 116 SMEs had joined this programme.

Over 1976–88, the total value of financial assistance by the Singapore government to SMEs amounted to S$1.5 billion, of which 88 per cent was in the Small Industries Financing Scheme. Grants of various kinds amounted to S$23.4 million and the Skills Development Fund S$48.6 million (Soon, 1994).

Public R&D

In 1991 the Singapore government launched a technology plan with an R&D target of 2 per cent of GDP by 1995 (in the early 1990s the figure was around 1 per cent). It selected a number of sectors for technology development,[36] and established a S$2 billion fund for R&D. The case of biotechnology is a good example of its approach to developing a domestic innovative base (Box 3.5). It illustrates how the government uses public R&D funds and institutions to build up basic research capability and so attract interest and R&D participation by MNCs, as part of a strategy of technological targeting.

Box 3.5   Singapore's Institute for Molecular and Cell Biology

A good example of the successful use of various government policies, institutions and financial instruments is Singapore's move into the scientific mainstream with the development of the Institute for Molecular and Cell Biology (IMCB). The IMCB is an ambitious project in the government's overall strategy to use high-technology to strengthen its economy. The government places this within the National Biotechnology Programme, started in 1988 to strengthen the national R&D base and fund biotechnology development. An important incentive under this programme is pioneer-industry status, which gives tax exemption for 5–10 years, with the largest benefits directed at technology-intensive and export-oriented projects. In addition, funding is provided by the government if there is active research collaboration with the public sector, with no specified limit to the available funding for R&D.

Supporting this effort is a strong push in basic research at the National University of Singapore (NUS), which houses the IMCB. The University conducts one-third of Singapore's R&D, and NUS scientists have made their mark in several areas including materials technology, microelectronics and information technology.

Singapore's decision to spend S$13.8 million to build IMCB and to provide annual funding of S$17.5 million was part of a broader approach to develop biotechnology, a field that fits the country's need (e.g., it requires few natural resources, has high value added, and can make strategic use of Singapore's global business networks). To nurture this industry, the EDB established Singapore Bio-Innovation (SBI) Pte Ltd, which by 1991 had invested S$41 million in 12 local biotech start-up firms with 1428 employees making health-care, food, and agricultural products. SBI also invests in overseas companies that might be strategic allies.

The investment in IMCB appears to be paying off scientifically. An IMCB group is at the forefront of research on tyrosine phosphates, a hot topic in cancer research. Another group is sequencing the genomes of several fish species, which could serve as a reference vertebrate genome for the human genome project. IMCB laboratories' innovative assay systems convinced Glaxo, the pharmaceutical MNC, to establish a S$31 million trust fund for a drug-screening centre within IMCB. Glaxo also invested S$30 million for a neurobiology lab focusing on genes that are expressed only in the brain.

Encouraged by these successes, the government expanded IMCB's research base by establishing the Bioscience Centre, which provides facilities for research at NUS and the Food Biotechnology Centre. The Bioprocessing Technology Unit, opened in 1990, seeks to improve purification, synthesis and fermentation methods for commercial production. The lab recently achieved large yields of TNF-[beta], which other companies, including Genzyme in the USA and Boehringer Mannheim in Germany, are keen to put into clinical cancer trials. The National University Medical Institute, being built near IMCB and the National University Hospital, is modelled on the US National Institutes of Health.

One obstacle to Singapore's quest for scientific success is its shortage of well-qualified scientists and engineers. To overcome this, the IMCB recruited for scientists from the West, offering them research freedom, ample funding and salaries of up to $50,000 for principal investigators. Those who accept IMCB's offer may qualify for renewable 3-year contracts. Singapore's own students represent the largest source of scientific talent at IMCB. Singapore's two polytechnics are training technicians to fill the growing demand from biotech labs and industries. In addition to tuition, graduate students at IMCB receive a handsome ($10,000 a year) stipend.

Source: Carroll (1994).

Taiwan

Taiwan's technology infrastructure for supporting its myriad SMEs is perhaps one of the best anywhere.[37] There are around 700,000 SMEs in Taiwan, accounting for 70 per cent of employment, 55 per cent of GNP and 62 per cent of total manufactured exports. The list of different efforts to assist them is impressive. In 1981 the government set up the Medium and Small Business Administration to support SME development and coordinate the several agencies that provided financial, management, accounting, technological and marketing assistance to SMEs. Financial assistance was provided by the Taiwan Medium Business Bank, the Bank of Taiwan, the Small and Medium Business Credit Guarantee Fund, and the Small Business Integrated Assistance Centre. Management and technology assistance was provided by the China Productivity Centre, the Industrial Technology Research Institute (ITRI) and a number of industrial technology centres (for metal industry, textiles, biotechnology, food, and information). The government covered up to 50–70 per cent of consultation fees for management and technical consultancy services for SMEs. The Medium and Small Business Administration established a fund for SME promotion of NT$10 billion. The 'Centre–Satellite Factory Promotion Programme' of the Ministry of Economic Affairs integrated smaller factories around a principal one, supported by vendor assistance and productivity-raising efforts. By 1989 there were 60 networks with 1186 satellite factories in operation, mainly in the electronics industry.

Several technology research institutes supported R&D in the private sector. The *China Textile Research Centre*, set up in 1959 to inspect exports, expanded to include training, quality systems, technology development and directly acquiring foreign technology. The *Metal Industries Development Centre* was set up in 1963 to work on practical development, testing and quality-control work in metal-working industries. It later established a CAD/CAM centre to provide training and software to firms in this industry. The *Precision Instrument Development Centre* fabricated instruments and promoted the instrument-manufacturing industry, and later moved into advanced areas like vacuum and electro-optics technology. The most important was perhaps the *Industrial Technology Research Institute* (ITRI).

ITRI conducted research and development for technology projects considered too risky. It had seven laboratories, dealing with chemicals, mechanical industries, electronics, energy and mining, materials research, measurement standards and electro-optics, but electronics was the institute's principal focus, with its Electronics Research & Service (ERSO) division accounting for two-thirds of the Institute's $450 million budget. ERSO has spun off laboratories as private companies including United Microelectronics Corporation (UMC) in 1979 and Taiwan Semiconductor Manufacturing Company (TSMC) in 1986, Taiwan's most successful integrated-circuit makers. The Institute for the Information Industry (III) was set up to complement ITRI's work on hardware by developing and introducing software technology.

Where the private sector was unable by itself to undertake complex or risky technologies, the government played a direct lead role. As noted above, the government (led on the technical side by ERSO) entered into a joint venture with Philips to set up the Taiwan Semiconductor Manufacturing Company, the first wafer-fabrication plant in the country. The government also strongly encouraged industry to contract research to universities, and half of the National Science Council's research grants (about $200 million per year) provided matching funds to industry for such contracts.

The Taiwan Handicraft Promotion Centre supported Taiwan's handicraft industries, particularly those with export potential. Its main clients were small entrepreneurs, most with under twenty employees. In addition, the Programme for the Promotion of Technology Transfer maintained close contact with foreign firms with leading-edge technologies in order to facilitate the transfer of those technologies to Taiwan.

The China Productivity Centre (CPC) promoted automation in industry to cope with rising wages and increasing needs for precision and quality. The CPC sent out teams of engineers, visiting plants throughout the country to demonstrate the best means of automation and solve relevant technical problems, at the rate of approximately 500 visits making some 2000 suggestions per year. CPC also carried out more than 500 research projects on improving production efficiency and linked enterprises to research centres to solve more complex technical problems.

The government set up a science town in Hsinchu, with 13,000 researchers in two universities, six national laboratories (including ITRI) and a huge technology institute, as well as some 150 companies specialising in electronics. The science town makes a special effort to attract start-ups and provides them with prefabricated factory space, five-year tax holidays and generous grants. In the 1980s the government invested US$500 million in Hsinchu.

Korea

The Korean government invested in a large array of technology infrastructure institutions. In 1966 it set up KIST (Korea Institute of Science and Technology) to conduct applied research of various kinds for industry. In its early years, KIST focused on solving simple problems of technology transfer and absorption. In the 1970s, the government set up other specialised research institutes related to machinery, metals, electronics, nuclear energy, resources, chemicals, telecommunications, standards, shipbuilding, marine sciences, and so on. These were largely spun off from KIST, and by the end of the decade there were 16 public R&D institutions. In 1981 the government decided to reduce their number and rationalise their operations. The existing institutes were merged into 9 under the supervision of the Ministry of Science and Technology. KIST was merged with KAIS (Korea Advanced Institute of Science) to become KAIST, but was separated again – as KIST – in 1989.

The government's strategic thrust in this sphere was mainly a series of *National R&D Projects* launched in 1982. These were large-scale projects which were regarded as too risky for industry to tackle alone but which were selected as being in the country's industrial interest. National Projects were conducted jointly by industry, public research institutes and the government, and covered activities like semiconductors, computers, fine chemicals, machinery, material science and plant-system engineering. 'Centres of Excellence' were formed in these fields to boost long-term competitiveness. National Projects were a continuation of the strategy of interventions to identify and develop the country's dynamic comparative advantage, orchestrating the different actors involved, underwriting a part of the risks, providing large financial grants, and filling in gaps that the market could not remedy (for data on the amounts see above, on technology financing).

Other policy measures to stimulate technological effort in Korea were more addressed to static market failures. These included the setting up of *Science Research Centres* and *Engineering Research Centres* at universities around the country to support R&D activities, the *common utilisation* of advanced R&D facilities by smaller private firms, and the construction of *science towns*. Daeduk Science Town has been under construction since 1974, and a large number of research and educational institutions are already well established there. The construction of Kwangju Science Town has started; others are planned. Technology *diffusion* was advanced by the Korea Institute for Economics and Technology, which collected, processed and disseminated scientific and technical information to industry.

Since the early 1980s a number of laws were passed to *promote SMEs*, leading to a perceptible rise in their share of economic activity (over 1975–86 the share of SMEs in employment, sales and value added rose by at least 25 per cent). This policy support was crucial to the reversal in their performance: it covered SME start-up, productivity improvement, technology development and export promotion. A host of tax incentives was provided to firms participating in these programmes, as well as finance at subsidised rates for using support services, credit guarantees, government procurement and the setting up of a specialised bank to finance SMEs. A number of other institutions were set up to help SMEs (such as the Small and Medium Industry Promotion Corporation to provide financial, technical and training assistance, and the Industrial Development Bank to provide finance). The government greatly increased its own budget contribution to the programme, though SMEs had to pay a part of the costs of most services provided to them.

To promote subcontracting to SMEs, the government enacted a law designating parts and components that had to be procured through them and not made in-house by large firms. By 1987 about 1200 items were so designated, involving 337 principal firms and some 2200 subcontractors, mainly in the machinery, electrical, electronic and ship-building fields. By this time, subcontracting accounted for about 43 per cent of manufacturing output and 65–77 per cent of the output values of the electrical, transport-equipment and other machinery industries. Generous financial and fiscal support was provided to subcontracting SMEs, to support their operations and process and product development. In addition, subcontracting SMEs were exempted from stamp tax and were granted tax deductions for a certain percentage of their investments in laboratory and inspection equipment and for the whole of their expenses for technical consultancy. Subcontracting promotion councils were set up by industrial subsectors and within the Korea

Federation of Small Business to help SMEs in the contractual relationship, to arbitrate disputes and monitor contract implementation.

India

It is worth looking at one aspect of the Indian technology-infrastructure policies, concerning the government's efforts to reform the public research institutes under the Council of Scientific and Industrial Research (CSIR). The role of these large and expensive institutions in improving industrial technology was seriously questioned in the 1980s. They had turned out to be largely 'ivory tower' institutes with few linkages to industry, producing technologies with little commercial application and doing research that industry had little use for. The government decided to reform the CSIR institutes to encourage greater cooperation with private industry.

In 1991 the World Bank launched an industrial technology-development project in India, with one component aimed at promoting industry-sponsored research at a number of public research institutes (PRIs) as well as the Indian Institutes of Technology, other universities and private research foundations. This component, the Sponsored R&D Promotion Fund, was initially allocated US$15 million, and was later allocated another $10 million and renamed the 'Sponsored Research and Development (SPREAD)' programme. The SPREAD component was aimed at promoting research awareness especially among small and medium-sized companies and changing the 'research culture' among the research laboratories and higher-education establishments to greater industry orientation.

The funds to finance the contracting of research were provided on a subsidised basis, at 6 per cent initially and 15 per cent subsequently, or with a royalty option. The finance was to cover up to 50 per cent of the cost of the research project contracted by industry, with the resources given as conditional loans (with eventual repayment at market rates if successful, and written off if not). The projects could cover prefeasibility studies, laboratory trials, prototype building and pilot plant operations for the development of new products and processes, significant improvements to existing products/processes and scaling up of a technology. The fund was administered by ICICI, a leading private-sector development finance company. The research projects were appraised by ICICI's Technology Group, and had to be carried out within two years. Firms receiving support had to show that their sponsorship was additional to what they were doing earlier.

By mid-1995, 53 firms had contracted 55 projects under the SPREAD programme, with an average project size of $400,000 and a loan component of $170,000 (42.5 per cent). So far, there had been no failures, but some 3–4 projects were likely to be cancelled. Most of the companies using the programme had never contracted research to a public research institute before; of the 53 firms, 23 were small, 22 medium and 10 large. Their activities were spread over pharmaceuticals, electrical and electronics, chemicals, machinery, metallurgy, automotive, biotechnology, food processing, paper, rubber and polymers. Some 60 different technology institutes were involved, including 16 Institutes of Technology/Science, 12 universities, 4 private research foundations, and 28 government laboratories. A broad range of new or improved technologies was developed, some fairly sophisticated. Overall, the project is highly successful, and the subsidy element has been minimal.

SOME CONCLUSIONS

This sketch of some technology policies in Asia shows the variety of experience and strategies that exists in the region. It is difficult to generalise across this range of economic, political, social and historical settings. Taking the evidence on technology policies in concert with the analysis of industrial policy in the NIEs presented elsewhere in this book, some broad conclusions can be drawn:

- Technology development faces a number of market failures, and governments need to mount interventions to overcome them and promote technology deepening and diversification. Such interventions can be successful in stimulating and guiding the nature of ITD.

- Technology policy has to be guided by a strategic 'vision' rather than being confined to countering market failures in a static sense. Because of the costs and risks involved in moving into more complex technologies and developing more advanced capabilities, there is a need for government support for such a vision. Since intervention resources are necessarily limited, this necessarily involves selectivity in assigning technological objectives, and a measure of infant-industry support to give strong signals to the private sector.

- An outward-oriented trade regime, that forces firms to compete in world markets and provides flows of technical information to them, is crucial to mounting interventions and overcoming the risk of secondary distortions that may result from infant-industry strategies. However, outward orientation does not mean a free-trade regime, and many successful strategies have combined selective import controls with export orientation.

- Within an outward-oriented regime, there are two broad options for technology upgrading: relatively *dependent*, based on 'internalised' means of technology transfer via FDI, and relatively *autonomous*, based on indigenous enterprises with 'externalised' technology transfer. The latter is more demanding in terms of policy and institutional support, but is likely to yield higher dividends in terms of the development of local technological and innovative capabilities. An FDI-dependent strategy can lead to considerable technological deepening and a competitive production base if the government adopts selective policies to stimulate upgrading of activities and supports this with investments in human capital of the appropriate type.

- The nature and intensity of technological support needed depends also on the industrial structure. Countries with relatively large numbers of SMEs have to provide greater R&D inputs and extension services than those with more concentrated structures and large firms. If they can, the former can develop competitive and flexible industrial structures based on specialised technologies. Countries that encourage large-sized domestic firms are able to enter 'heavier' areas of technological activity, because of the minimum economies of scale in such activity, the risks involved and the need to internalise deficient capital and other markets.[38] However, this is a very risky and potentially costly strategy, and imposes costs in terms of high levels of industrial concentration; it needs extremely close monitoring and the ability to discipline poor performers if such a strategy is to succeed.

- SMEs have special need of technology support. The most effective approaches involve extension services delivering comprehensive 'packages' of assistance comprising technical know-how, finance, management skills, training and sales information. Such delivery is a skill-intensive task, involving a strongly proactive approach by the services concerned. Extension services need an element of subsidy, since SMEs are unable or unwilling to pay full market costs for a considerable period of time.

- Investments in skill creation can be both strategic and functional; both are needed as the industrial structure becomes more complex.

Both call for prolonged, intense and complex institutional development.

• The financing of technological activity has to be improved as economies mature industrially and technology-based firms start to appear. Technology financing can be used to successfully support entry into and upgrading of targeted activities.

• The technology infrastructure plays a crucial role in supporting industrial technology, and can be used to promote the development of new technologies. However, given the danger of creating 'hothouse' technologies with little industrial use, countries must ensure that research institutions become closely tied to industrial needs.

REFERENCES

Amsden, A. (1989), *Asia's Next Giant* (Oxford: Oxford University Press).
Brautigam, D. (1995), 'The State as Agent: Industrial development in Taiwan, 1952–1972', in H. Stein (ed.), *Asian Industrialisation and Africa* (London: Macmillan).
Carroll, A. M. (1994), 'Technology Development Experiences of Japan, Singapore, Korea, Taiwan, and Hong Kong' (Washington, DC, draft).
Clifford, M. (1994), 'Trading Up', *Far Eastern Economic Review*, 26 May, pp. 68–9.
Dahlman C. J. and Sananikone, O. (1990), 'Technology Strategy in the Economy of Taiwan: Exploiting foreign linkages and investing in local capability' (World Bank, draft).
Enos, J. and Park. W. H. (1987), *The Adaptation and Diffusion of Imported Technologies in the Case of Korea* (London: Croom Helm).
Gilley, B. (1995), 'New Model', *Far Eastern Economic Review*, 21 December pp. 50–2.
Hobday, M. G. (1995), *Innovation in East Asia: The Challenge to Japan* (Cheltenham: Edward Elgar).
Hou, Chi-Ming and Gee, S. (1993), 'National Systems Supporting Technical Advance in Industry: The case of Taiwan', in R. Nelson (ed.), *National Innovation Systems* (Oxford: Oxford University Press).
Kim, L. (1993), 'National System of Industrial Innovation: Dynamics of capability building in Korea', in R. R. Nelson (ed.), *National Innovation Systems: A Comparative Analysis* (Oxford: Oxford University Press).
Lall, S. *et al.* (1994), *Malaysia's Export Performance and its Sustainability* (Manila: Asian Development Bank, draft).
Lall, S. (1987), *Learning to Industrialize* (London: Macmillan).
Lall, S. (1990), *Building Industrial Competitiveness in Developing Countries* (Paris: OECD).

Mowery, D. C. and Rosenberg, N. (1989), *Technology and the Pursuit of Economic Growth* (Cambridge: Cambridge University Press).

Najmabadi, F. and Lall, S. (1995), *Developing Industrial Technology: Lessons for Policy and Practice* (World Bank, OED Study).

OTA (1990), *Making Things Better: Competing in Manufacturing* (Office of Technology Assessment, Washington, DC: US Senate).

Pack, H. and L. E. Westphal (1986), 'Industrial Strategy and Technological Change: Theory versus reality', *Journal of Development Economics*, 22(1): 87–128.

Rhee, Y. W., Ross-Larson, B. and Pursell, G. (1984), *Korea's Competitive Edge* (Baltimore: Johns Hopkins Press).

Selvaratnam, V. (1994), 'Innovations in Higher Education: Singapore at the competitive edge', (Washington, DC: World Bank), Technical Paper No. 222.

Song, J.-H., (1995), 'Technology Financing in Korea', *The Korea Development Bank, Economic and Industrial Focus*, December, pp. 1–10.

Soon, Teck-Wong (1994), 'Singapore', in S. D. Meyanathan, *Industrial Structures and the Development of Small and Medium Enterprise Linkages: Examples from East Asia* (World Bank, Economic Development Institute).

Soon, Teck-Wong and Tan, C. S. (1993), 'Singapore: Public Policy and Economic Development' (World Bank), background paper for *The East Asian Miracle*.

Stiglitz, J. E. (1989), 'Markets, Market Failures and Development', *American Economic Review Papers and Proceedings*, 79(2): 197–202.

Stoneman, P. (1987), *The Economic Analysis of Technology Policy* (Oxford: Oxford University Press).

Teubal, M. (forthcoming), 'R&D and Technology Policy in NICs as Learning Processes', *World Development.*

Wade, R. (1990), *Governing the Market* (Princeton: Princeton University Press).

Westphal, L. E. (1990), 'Industrial Policy in an Export-Propelled Economy: Lessons from South Korea's experience', *Journal of Economic Perspectives*, 4(3): 41–59.

Wong, Poh-Kam (1995), 'Singapore's Technology Strategy', in Denis F. Simon (ed.), *The Emerging Technological Trajectory of the Pacific Rim* (New York: East Gate), pp. 103–31.

World Bank (1993), *The East Asian Miracle: Economic Growth and Public Policy* (New York: Oxford University Press).

Young, A. (1991), 'Learning by Doing and the Dynamic Effects of International Trade', *Quarterly Journal of Economics*, CVI(2): 369–405.

4 'The East Asian Miracle' Study: Does the Bell Toll for Industrial Strategy?

INTRODUCTION

The World Bank's *East Asian Miracle* study[1] has been anticipated with considerable interest by the development community. It was intended (not least by the Japanese government that financed it) to be an objective re-examination of the role of government interventions in economic, particularly industrial, development. It reflected a widespread unease that the Bank was too strongly committed to a neoliberal view of the development process.[2] This approach evolved over the 1980s, drawing mainly upon evidence from East Asia, and fuelled by a shift in mainstream economics and political perceptions in the leading developed countries. It formed the basis of the Bank's subsequent lending and policy advice, and was at the core of the structural adjustment programs that have shaped policy in many developing countries. This review confines itself to the study's analysis of *industrial policy*, concentrating on the approach and the analysis of the established NIEs (the 'Four Tigers').

The re-examination of the East Asian 'miracle' is significant for many reasons. It is relevant for development research and analysis, since the World Bank is the foremost institution involved in such work. It is important to the governments that are trying to emulate the high-performing Asian economies (HPAEs), in particular the leading Asian NIEs, Korea and Taiwan, as well as to those that are in the throes of liberalisation. For the Bank, it is central to its approach to policy reform in the developing world (and economies in transition). It may even be important for the developed world, since Japan is one of the countries in the study, and industrial policy remains a live issue.

The theory and evidence that supported the Bank's neoliberal approach have been increasingly questioned in recent years.[3] One of the challenges was by the Bank's own Operations Evaluation Department (OED). In its (1992) study of industrialisation in Korea, India and

Indonesia, the OED acknowledged that much of government intervention, especially in import-substituting countries like India, had been costly and inefficient. However, the evidence on Korea (drawn from Bank reports) suggested that selective interventions had been critical to the pace and content of its industrial development. The OED concluded that certain forms of selective interventions may be necessary to overcome market failures and promote technological deepening and entry into complex technologies. These interventions differed radically from those in 'classic' import-substitution, but it was possible for governments to design and implement them if they had the right objectives, skills and information, and if they set the interventions in a framework of export orientation. This study may well have been instrumental in inducing the Bank to launch the present study.[4] In this institutional context, a consideration and amplification of the OED arguments was perhaps one of the *Miracle* study's objectives.

The *Miracle* study has many strengths. Its main contribution may be that it admits the extent and pervasiveness of interventions in most HPAEs. The facts are generally well known to those familiar with the region, but the Bank's official endorsement, in a highly publicised document, is a major step. The reaction that this evokes in some quarters is itself indicative of the significance of this endorsement and of the institution's image as a defender of the neoliberal faith.[5] Some of the interpretation of the evidence is interesting. Several of the policy implications drawn are sound: the need for 'fundamentals' like sound macroeconomic management, export orientation and more education. For those who like statistical exercises, there are many regressions. There are useful insights into how interventions were managed in the larger NIEs and Japan, how the bureaucracy was insulated and how government efficiency may be improved.

In essence, however, the *Miracle* study is a flawed work. The presentation of empirical evidence is selective and incomplete, and its interpretation sometimes tendentious and biased. The main policy conclusions on getting the 'fundamentals' right are neither original nor debatable: no one proposes bad macro-policies; the trade-strategy debate is no longer alive – the debate is now on the role of states versus markets in successful export-oriented economies; and the need for primary schooling is hardly in dispute. The more interesting questions concerning the kind of higher education and specialised training that may be conducive to industrial development are not explored. While some of the statistical exercises are of interest, especially to establish the correlation between education and growth, others (like the ones on total factor-productivity growth) are subject to many measurement and interpretation problems. The analysis of government capabilities stops short of drawing implications for helping other governments to mount desirable strategies.

The main reservations are with respect to its treatment of *industrial policy*, the focus of this review. The most controversial part of the study, and the one most debate centres around, is the role of selective (crudely put as 'picking winners') as opposed to functional interventions. Had the study found that selectivity was necessary for, or even conducive to, industrialisation of the sort witnessed in East Asia, the foundations of much of its recent policy work and adjustment programmes would have been shaken. The implications for the Bank would have been enormous.

However, no such embarrassment is posed. The study concludes that some interventions, especially in capital markets, may have helped some of the larger, leading industrialisers (Japan, Korea and Taiwan). However, selective promotion in general was not really effective in meeting its objectives, cannot be undertaken by other governments without the skills and impartiality of the East Asians, and is not in any case an open policy option now. Thus, it ends up with soothing noises for the Bank: governments should only undertake 'market-friendly' interventions and get economic 'fundamentals' right.

Is it then the final word on industrial policy? Should the debate be wound up and neoliberal prescriptions endorsed? Not on the basis of the *Miracle* study. The evidence presented, while revealing, is not complete. The definitions of market failure and industrial policy used are biased and partial. The theoretical framework for reviewing the effects of industrial policy is inadequate. Some logical implications of the arguments are not followed through. The policy conclusions on industrial policy are unwarranted and misleading. It fails to tackle, even acknowledge, some of the critical arguments of the OED report. Despite its many contributions, therefore, the study is pusillanimous and disappointing.

THE 'MARKET-FRIENDLY' METHODOLOGY

Let us start with methodology. The study poses three alternative views of the role of policy in the success of the HPAEs. The 'neoclassical' view is that East Asian success was due to neutral incentives between domestic and foreign markets and limited government interventions, permitting the realisation of static comparative advantage.[6] The 're-visionist' view is that interventions, including selective ones, were central to East Asian success because of the presence of pervasive market failures.[7] The 'market-friendly' view is that growth resulted from functional interventions in 'fundamentals' like promoting education, stable macro-management, competitive regimes and openness to trade. Selective (and so presumably 'market-unfriendly') interventions were generally unsuccessful and governments generally failed in exercising selectivity.[8] This approach is the one endorsed by the study, and so has to be treated at some length here.

The study convincingly disposes of the neoclassical interpretation on empirical grounds, though the theoretical underpinnings of this approach are not tackled.[9] It establishes that interventions were pervasive in most HPAEs. All intervened functionally, and most also intervened selectively: in imports, export promotion, credit allocation, technology flows, firm growth, foreign investment and public ownership. However, there were large differences between them in the pattern and extent of interventions. Japan and Korea led in the scope and detail of interventions, with Taiwan close behind. Among the smaller Tigers, Singapore intervened pervasively and selectively, but did not protect domestic industry. Hong Kong intervened the least selectively, though it provided the 'fundamentals'. While growth was clearly possible without selective interventions, the study agrees that this does not dispose of the revisionist case. Since there are many patterns of industrial development and many possible paths to success, the question still remains whether selective interventions helped the economies that practised them to develop the kinds of industrial structures and resources they wanted, and whether market-friendly policies would have been as efficient. The rest of the review seeks to evaluate how the *Miracle* study poses and answers this question.

In the standard neoclassical approach all markets are assumed efficient. Product markets give the correct signals for investment in new activities and factor markets respond to these signals. At the firm level, given perfect competition, information, foresight and efficient factor markets, the optimum point on the production possibility frontier is chosen according to prevailing factor prices. All firms are by definition equally efficient: technology is freely available, with full knowledge on techniques available to all firms – most importantly, it is costlessly and instantly absorbed, and any 'learning' process is known, predictable and automatic. Over time, as factor prices change to reflect changing endowments, their activities change accordingly – this represents the optimal pattern of specialisation and forms the basis for evolving comparative advantage. With these assumptions, it follows logically that interventions can only be distortionary. There is, however, nothing in neoclassical theory which says that if the assumptions are changed and market imperfections admitted, its welfare and policy conclusions remain the same. The extent to which the initial assumptions apply in practice is an empirical question. It is neoclassical development economists who, generally implicitly, have tended to assume that markets in developing countries are in fact efficient, and that imperfections are of little practical or policy significance.

The market-friendly approach (as represented by the study) drops some of the assumptions of the neoclassical approach. It accepts that factor markets may not operate perfectly, and that education markets in particular may need interventions to create the human-capital base for industrialisation. These interventions are taken to be 'market-friendly' (i.e. non-selective), on the implicit assumption that skills are generic and fungible. Thus, different patterns of industrialisation are taken to have similar skill needs; all HPAEs invested in general education and no selectivity was involved. There was thus no need for selective interventions to coordinate skill and industrial markets. This is questionable, but the otherwise useful analysis of human capital in the study does not explore this issue.

The market-friendly approach also differs from the neoclassical development approach in accepting that there may be market failures in coordinating investment decisions within industry for several reasons: missing information markets,[10] capital-market deficiencies, economies of scale, interdependent investments in vertically related activities, externalities in skill creation and learning, and 'multiple linkages'.[11] Since many of these failures differ in incidence and intensity across different activities, interventions to correct for them necessarily have to be selective. In their absence, theory predicts that resource allocation would be suboptimal and growth constrained.

There are two features of the way the *Miracle* study deals with the issue of selectivity which are central to its market-friendly conclusions on industrial policy. The first is its treatment of the market failures it does mention. Having acknowledged the theoretical case for selectivity, the study should have analysed at length how important the relevant market failures were and how they were addressed in different HPAEs. While the study notes that different governments had different selective policies, and supplemented information flows by setting up mechanisms for business and government to interact, it makes no attempt to assess what market failures these different policies were addressing, how well they succeeded and what might have happened had they not intervened. Did these market failures really exist in practice? What was the possible impact on industrial development if they were not remedied? Were they in fact remedied by the selective interventions mentioned? If they were, even partially, does this provide a case for mounting and refining industrial policy? The study evades these questions, giving the impression that simply by mentioning the possibility of market failures and the desirability of government–business interaction it has dealt with the issue.

The second problem is that the list of market failures is incomplete. As argued below, there are several market failures that confront developing-country firms attempting to industrialise. The nature of these failures is not general, given with reference to a static equilibrium model that applies to all countries. They depend on the *specific objectives of the country concerned*, in terms of the activities it wants to enter, the extent of local integration it aims at, and the level of indigenous technological competence and innovative capabilities it wishes to develop. The relevant intervention needs have to be assessed in this context.

Such an assessment of market failures calls for an examination of *technological learning at the micro-level* (see below). The study in fact uses an oversimplified approach that assumes costless and automatic learning and upgrading of industrial technologies. Furthermore, there were crucial *differences in technological objectives* between each of the Tigers and Japan. These strategies led to widely differing industrial structures, export specialisation, degrees of local content, indigenous technological capabilities, reliance on different modes of technology transfer and the continuing involvement of government. These differences, central to the assessment of their industrial strategies, are ignored in the *Miracle* study. This allows it to present a biased and misleading evaluation of the effects of industrial policies.

Finally, the distinction drawn between 'market-friendly' and selective interventions is spurious. There is no economic basis to distinguish between functional and selective interventions: any intervention that corrects for market failures is market-friendly. Nor does economics provide any *a priori* reason for arguing that beyond functional interventions 'governments are likely to do more harm than good'.[12] This is a political statement of dubious empirical value – many governments intervened badly in the past but others intervened very efficiently. This suggests that there may be ways in which the quality of interventions can be greatly improved (see final section, below).

The 'revisionist' approach is tackled obliquely by the study. There is clearly no theoretical difference between revisionist and other approaches. The only difference lies in their reading of market failures and of the ability of governments to intervene. The study can only counter the revisionist conclusion by arguing that market failures were in fact unimportant, and selective interventions not effective. This it does, as shown below, by a selective use of the evidence and by evading some of the most significant issues.

A DIFFERENT PERSPECTIVE ON MARKET FAILURES

It may be useful to describe briefly the micro-level perspective on industrial development.[13] This provides a more realistic and complete framework for the analysis of market failures and the need for interventions than the simple neoclassical model based on a unique static equilibrium. Industrial success in developing countries depends essentially on how well individual firms manage the process of technological and managerial development. Technological development does not mean innovating new technologies but, at least at the start, using imported technologies efficiently. Technology is not perfectly transferable like a physical product: it has many 'tacit' elements that need the buyer to invest in developing new skills and technical and organisational information.

The process of gaining technological competence is not instantaneous, costless or automatic, even if the technology is well diffused elsewhere. It is risky and unpredictable, and often itself has to be learnt: in developing countries new firms may not even know what their deficiencies are or how to go about remedying them. The development of competitive capabilities may be costly and prolonged, depending on the complexity and scale of the technology. It involves interactions with other firms and institutions: apart from physical inputs, it calls for various new skills from the education system and training institutes, technical information and services, contract research facilities, interactions with equipment suppliers and consultants, standards bodies, and so on. The setting up of this dense network of cooperation needs the development of special skills. This constant and uncertain process of learning differs radically from the standard neoclassical model of firm development, and leads to different policy implications.

Industrial development is not just about starting new activities. As economies progress and mature, it involves '*deepening*' in any or all of four forms – technological upgrading of products and processes within industries, entry into more complex and demanding new activities, increasing local content, and mastering more complex technological tasks within industries (from those relevant to assembly to those needed for more value-added activity, adaptation, improvement, and finally design, development and innovation). Each involves its own learning costs. These costs differ by activity, rising with the sophistication of the technology, the extent of linkages and the level of technological capabilities aimed at. Progressive deepening is to some extent a natural part of industrial development, but it is not inevitable. Its pattern and incidence differ greatly, *depending on the strategies pursued by the government.*

The process of capability development may face various market failures. Free markets may not give correct signals to resource allocation, first, between simple and difficult activities and, secondly, between physical investments, technology purchase and internal technological effort. The first is the basis of the classic case for infant-industry protection. In the presence of learning costs, a latecomer to industry necessarily faces a disadvantage compared with those that have undergone the learning process.[14] Given the unpredictability, lack of information, and capital-markets imperfections that are endemic to developing countries, exposure to full import competition can prevent entry into activities with relatively difficult technologies.

Since learning costs differ between activities, interventions to ensure efficient resource allocation have to be selective rather than uniform. In simple activities the need for protection may be minimal, because the learning period is relatively brief, easy to get information on, and predictable. In fact, in labour-intensive activities like garment assembly, the wage cost advantage of developing countries may offset the learning costs completely, making protection unnecessary. In complex activities, with large scales, advanced information and skill needs, wide linkages and intricate organisations, by contrast, the learning process could spread over years, even decades.[15] These may never be undertaken (unless there is a strong natural-resource cost advantage) unless protection is given.

The second kind of allocation, affecting the deepening of capabilities *within* activities, can also suffer market failures. Arrow (1962) noted long ago that a free market may fail to ensure optimal innovative activity because of imperfect appropriability of information and skills. However, developing countries face an additional problem. It is generally easier to import foreign technologies fully 'packaged', where the process is commercially proven, and the supplier provides the hard and software, does the start-up, training and adaptation, and manages the operation and marketing. In its extreme form, fully 'internalised' technology transfer takes the form of wholly foreign-owned direct investment. This is an effective and relatively less risky way to access technology, but it leads to little capability acquisition in the developing country apart from production skills. The move from production to innovative activity involves a distinct strategic decision which foreign investors tend to be unwilling to take in developing countries. There is generally less investment in design, development and innovation by foreign compared with local firms, and the externalities generated by the technological activity that does take place tend to be captured by the investor rather than the host economy.

There is, in other words, a risk of market failure in capability deepening because of the learning costs, very similar in nature to infant-industry considerations. To ensure socially optimal allocation, it may be necessary to selectively restrict technology imports in 'internalised' forms and promote those in 'externalised' forms (licensing or equipment): here the buyer has to do much more 'homework' and so can develop broader and deeper capabilities. However, there are many technologies that are only available through direct investment, or that are too complex for local capabilities – these have to be imported in internalised forms. In order to bear the risks, costs and expenses of locally absorbing and very complex technologies, it may also be necessary to promote large firm size. Technological deepening can be a legitimate goal of industrial policy, since the development of indigenous design and innovative capabilities has many beneficial externalities.

These considerations also apply to deepening in the sense of increasing local integration: the development of local suppliers and subcontractors. Apart from the production benefits, these linkages speed the diffusion of technology, increase specialisation and raise industrial flexibility. In particular, the development of local equipment suppliers can raise the generation and diffusion of technology.[16] Because of these externalities, there may be a valid case for promoting *entire sets* of related activities, which would not otherwise be able to coordinate their investments. There may also be a case for selecting those sets of activities that offer higher learning potential, because of the advanced technologies involved.[17] Again, the nature of the market failures depends on the economy concerned and its technological ambitions.

Since the skills and information needs of different industrial activities differ, interventions in these factor markets have to be *integrated* to interventions to promote activities or technological deepening. If the government targets, say, the electronics industry for promotion, it has to ensure that electronics skills are provided by universities, technical information and support facilities on electronics are provided by infrastructure institutions, and so on. 'Market-friendly' interventions are thus necessarily selective once skills and information become specific rather than generic. Finally, since protection reduces the incentive to invest in capability development, industrial policy must provide *offsetting incentives* in the form of performance requirements. This could take several forms, but the easiest and most powerful is probably the need to enter export markets in a short period.

To recount, the promotion of industrial development may need interventions to overcome market failures in resource allocation between activities and within firms. These interventions have to be selective and geared to the learning process within firms. They may cover particular activities or sets of activities, and they may require the promotion of large firms. They have to be integrated with selective interventions in factor markets, including measures affecting the mode of technology import. They have to be offset by measures that provide incentives to invest in capability development.

It is immediately obvious that the aims of industrial policy are not simply to move from labour-intensive to capital-intensive activities. A two-factor neoclassical model does not begin to capture the complex processes of upgrading and learning involved. The success of policies aimed at deepening in the forms mentioned can only be assessed at a very detailed level, and only in the context of the objectives of the governments imposing those policies. The Tigers provide ample material for this investigation.

Hong Kong, for instance, had few strategic objectives beyond the preservation of a liberal trade and financial centre. It therefore intervened the least, to provide export information, general education, some selective skills (the textile and garment design centre) and some technological support (the Productivity Centre, see Chapter 3). However, it had behind it a century of experience of trade and finance, a large presence of expatriate British trading and financial enterprises (which, over time, transmitted their skills and information to locals), and an influx of textile entrepreneurs and technicians from Shanghai who had already undergone a learning process. These unique advantages (which incidentally make it absurd to draw policy generalisations from Hong

Kong for other small, open, developing economies) allowed it to successfully enter, and upgrade within, light industry, but not to deepen its industrial or technological structure.

By contrast, Korea started with light industry, but from the start protected and intervened in various other ways to change this structure. It protected and directed domestic (often subsidised) credit to promote entry into complex technologies, raised local content, restricted foreign direct investments, created giant conglomerates, directed and promoted R&D, selectively intervened in skill formation, and set up a comprehensive science and technology infrastructure geared to the needs of its selected industries. All these measures formed a *coherent package*, aimed at the composite strategic objective of entering difficult industries, with significant local integration, primarily under national ownership, and with a steady upgrading of local innovative capabilities.[18] Each element was necessary to this strategic objective. The other Tigers had different objectives somewhere between these two, and adopted consistent strategies to achieve them.

The success of the NIEs' industrial policies has to be judged in the context of these different strategies. The nature of what governments regarded as market failures depended on their objectives: what might have been considered satisfactory progress in Hong Kong may have been regarded as inadequate (and so prone to market failure) in Korea. The effects of industrial policies cannot therefore be assessed without reference to strategies pursued – the use of a static theoretical optimum to assess market failures misses the whole point. A proper assessment on industrial policies would involve, among other things, a detailed examination for the NIEs of the complexity of industrial activities within and across industries, the composition and technological sophistication of their manufactured exports, the extent of local integration, and the development of indigenous technological deepening. These were the facets of industrialisation in which their strategies differed, for which market failures were identified and to which interventions were addressed. Let us return to the *Miracle* study and see what it does.

THE *MIRACLE* STUDY'S EVALUATION OF INDUSTRIAL POLICY

The evaluation of industrial strategy is in chapter 6, in particular in the section on 'Using the International Market: Trade and Industrial

Policy'. The study acknowledges that some market failures exist, but the list is derived essentially from a static neoclassical model.[19] It ignores those that arise in firm-level learning, particularly those that face industrial deepening in the sense noted above. As such, it does not consider the strategic industrial-policy objectives pursued by the NIEs and Japan. It describes many (but not all) of the interventions undertaken, but makes no attempt to assess whether they were overcome, or what would have happened had interventions not been made. Where the achievements of industrial policy are measured, the definition of 'industrial policy' is tilted to produce a particular result. The description and analysis of interventions are biased to suggest that it was conformance to free markets that led to industrial success. This is done in several ingenious ways.

First comes a section on 'The Closeness of East Asian Domestic Prices to International Prices', which argues that HPAEs distorted their prices less than other developing countries. This may be soothing to the Bank, but is not really relevant – the real issue is not whether import-substituting economies intervened *more* than the HPAEs (since 'classic' import substitution is now universally accepted as inefficient), but whether the export-oriented East Asian economies distorted their prices at all, and if this was *desirable for industrialisation*. This issue is skated over. The study then goes on to say that 'Japan, Korea and Taiwan, China, rank in the fifth and sixth deciles, below such developing economy comparators as Mexico, Brazil, India, Pakistan and Venezuela. This is consistent with the evidence that the three northern HPAEs intervened far more frequently and systematically than the south-eastern HPAEs.'[20] This has two important implications, neither of which is emphasised by the study. First, there is a clear positive relationship *within* the group of HPAEs between the extent of industrial deepening and the degree of price distortions: the most successful industrialisers in Asia (and arguably in the world in recent experience) distorted their prices significantly. Secondly, these successful industrialisers distorted their prices *more* than several less successful import-substituting economies. Thus, it was not conformance to world prices that led to their success, but the discipline of export orientation. This conforms to the predictions of the capability approach, but blows the 'market-friendly' thesis out of the water.

Then comes a section on 'Openness to Foreign Technology'. Of course the East Asians were open to foreign technology, as were most other countries in Asia and Latin America (with the exception of India[21]). However, there is no analysis in the study of the effect of *different modes* of technology import on the industrial development: different forms of openness were perceived by these countries to have different implications for the development of their own innovative capabilities, and this formed a crucial element of their industrial strategies. Japan, Korea and Taiwan – the countries with the deepest local technological capabilities – were selective on, even hostile to, foreign investment; the study notes this in passing but draws no lessons for industrial policy. The causal connection between protection and selectivity in FDI, indigenous capability development and large domestic firm size (in Japan and Korea) seems clear. But it goes against the 'market-friendly' theology and so is not mentioned.

In this context, the study also fails to analyse the striking differences in *local technological effort* (in particular, research and development) achieved by the HPAEs. Clearly, this was crucial to the ability of countries like Japan, Korea, and, to a lesser extent, Taiwan, to absorb and build upon imported technologies, and to their international competitiveness and export growth.[22] As a consequence of this neglect, the study also fails to mention the widespread interventions used (in science institutions, by national technology strategies, by technical extension and support services, and by various incentives and subsidies at the firm level) to achieve this increase. The use of 'openness' serves to evade the real issues of industrial policy in this area.[23]

The 'bottom line' of the study is the section entitled 'Industrial Policies'. This defines what the study believes industrial policy's objectives are, then develops measures to test whether these objectives were achieved. This is perhaps the crux of the whole study – the final test of the role of government in industrial development – and so merits special attention. Industrial strategy is defined as *'government efforts to alter industrial structure to promote productivity-based growth'*.[24] This seems innocent enough, but its application shows that it is in fact slanted.

Industrial policy essentially comprises all actions taken to promote industrial development beyond that permitted by free market forces. In the larger East Asian NIEs and Japan its basic thrust was to overcome market failures that deterred deepening: upgrading into more complex products and processes within existing activities as well as entering into advanced new activities, raising local content, and developing the base of local technological capabilities. This encompassed much more than 'altering industrial structure', though in the long run some structural effects were bound to occur. However, many effects occurred *within* existing activities as well as across them. Many involved interventions in related factor markets such as education and training and the technology infrastructure to provide the specific inputs needed for the promoted technologies. Their effectiveness can therefore only be evaluated by looking at the range of interventions in all these markets to promote specific technologies.

How does the study handle this task? It uses two empirical tests to evaluate the effectiveness of the study's version of industrial strategy. Neither comes remotely near to assessing strategies as they were conceived and implemented by the governments concerned. First, the study looks at whether industrial policy altered industrial structure in comparison to what changing factor intensities and prices predict. This uses a simple two-factor neoclassical model to predict structure. As noted earlier, a two-factor approach is totally inappropriate to the nature of the learning and deepening process that industrial policy aimed at. The model should, in other words, have included several more factors like technology, specialised skills and so on, but it is not clear how these could have been quantified.

Two comparisons are made, with international norms for shares of particular industries, and for changes over time for the Tigers and Japan. Both are at the *two-digit level of aggregation*, which is so broad that it cannot possibly capture the deepening and upgrading process. The comparisons with international norms may in any case be misleading because they reflect the promotion efforts of other countries in the sample. Changes caused by industrial policies *within* particular countries simply cannot be measured at this level. Since industrial policy was not intended to suppress existing activities, and some policies tried to upgrade technologies within mature sectors, moreover, changes in the broad balance of activity cannot be taken to reflect their effectiveness. In any case, the gestation period involved in mastering new technologies in heavy industries is relatively long, and the study does not take a sufficient period to account for this.[25]

The conclusion of the study, not surprisingly, is that industrial policy had little structural effect. This flies in the face of overwhelming micro-level evidence that most promoted activities would not otherwise have been undertaken. More surprising is the total lack of any effort to use *other indicators* of policy effectiveness that take the countries' own objectives into account. Despite all the statistical work, no attempt is made to look at the rate of growth of exports by country,[26] the skill or technology composition of exports and production, indices of local technological development, local capital goods and engineering, R&D, and so on. The whole exercise is a masterly piece of obfuscation.

The second test is whether promoted sectors had higher TFP growth than others. It predictably finds that they did not. Apart from the inherent difficulties in measuring and interpreting TFP,[27] it is not clear what this test shows at such a high level of aggregation.[28] It cannot capture the productivity effects of upgrading within activities or across technologically distinct activities. In any case, it is misleading even on its terms: different activities start at different distances from the frontier, and higher TFP figures may simply show that the sectors concerned were behind others at the start. Differences in TFP may also reflect different periods of maturity of different activities.

The chapter goes on to argue that export activity generates learning and productivity growth. This is true to a great extent, but it ignores the fact that new export activities in the larger Tigers and Japan were creatures of industrial policy. Export orientation provided the discipline and incentives that allowed *other* interventions to succeed, and mere export-orientation by itself would not have allowed many of these activities to be set up. While a different part of the study rightly makes much of the export-based 'contests' that provided the discipline for selective interventions, the final assessment ignores the true contribution of export orientation to industrialisation – the right framework for selective interventions.

POLICY IMPLICATIONS OF THE STUDY

The final chapter, drawing the policy implications, wraps up the findings of the study with several sensible comments. Since no one disagrees with the need to have sound macro-management and better education, however, there is little value added here. Where value *is* added, the study is not entirely objective. The conclusion that 'selective interventions were neither as important as their advocates suggest nor as irrelevant as their critics contend' is not justified, since the positive contribution of selective interventions has not been properly analysed. The assertion that, in the Tigers, 'industrial policy did not alter the industrial structure or patterns of productivity change' is tendentious. The study does admit that some selective interventions in capital markets *may* have succeeded in Japan and Korea, but this is not proved by the empirical evidence it provides. More importantly, it looks at each intervention separately rather than as part of an overall strategy, and so it is completely unable to evaluate its real contribution. Credit-market interventions, for instance, could not have worked if they had not been integrated with protection, technological promotion, selective skill creation, and so on.

One of the most useful contributions of the study is its description of how the risks of intervention were contained, information was exchanged between business and government, and the civil service strengthened and insulated. For those who favour selective industrial policy, this is useful for helping governments to intervene more effectively. The study avoids this obvious conclusion by playing down the market failures that call for selectivity – all institutional strengthening is then to be directed to market-friendly policies – and by arguing that circumstances in Japan, Korea and Taiwan were historically unique. Certainly there may be some elements of these countries that are not reproducible, but several critical elements, like protection in the context of export orientation, selective education and training, the promotion of indigenous technological effort, or the creation of an efficient civil service, are. It is the function of such studies, and of international institutions like the Bank, to help with desirable selective interventions by appropriate advice and information (as was called for in the OED study), rather than strive to deny the case for such interventions. Certainly the market failures that developing countries face in industrialisation are not unique to the East Asian economies, and adopting 'market-friendly' policies fails to address many of them.

The final argument against industrial policy in the study is that the international scene, GATT, and the pressures exerted by the developed Western countries, are inimical to selective interventions. This is certainly true. Many of the instruments of industrial policy are increasingly constrained in the name of liberalisation. However, if a valid economic case for intervention exists, surely the study should ask for a review and easing of those pressures? And surely the Bank should use its considerable influence to favour industrial development in the weaker countries? This is evidently out of bounds, since the Bank itself is among the leading forces in imposing sweeping liberalisation and 'market-friendly' solutions.

The final lessons of East Asia drawn by the study for the rest of the developing world are, therefore, tame and partisan. They reflect neither theory nor evidence. In the final analysis, the real lessons of industrial development in East Asia are still to be drawn by the Bank. To return then to the subheading: Does the *Miracle* study toll the bell for industrial strategy? No, the bell merely signals another round in the ring.

REFERENCES

Amsden, Alice H., *Asia's Next Giant: South Korea and Late Industrialization* (New York: Oxford University Press, 1989).

Arrow, K., 'Economic Welfare and the Allocation of Resources for Innovation', in R. Nelson (ed.), *The Rate and Direction of Innovative Activity* (Princeton: Princeton University Press, 1962), pp. 609–26.

Economist, 'Message in a Miracle' and 'Economic Miracle or Myth?' (2–8 October 1993), pp. 18–19 and pp. 73–4.

Far Eastern Economic Review, 'Question of Faith: Japan challenges World Bank orthodoxy', (12 March 1992), p. 49.

Jacobsson, S., 'The Length of the Learning Period: Evidence from the Korean engineering industry', *World Development*, vol. 21, no. 3 (March 1993), pp. 407–20.

Lall, S., 'Understanding Technology Development', *Development and Change*, vol. 24, no. 4 (October 1993), pp. 719–53.

Lall, S., 'Technological Capabilities and Industrialization', *World Development*, vol. 20, no. 2 (February 1992), pp. 165–86.

Lall, S., *Building Industrial Competitiveness in Developing Countries* (Paris: OECD Development Centre, 1990).

Lucas, R. E., 'On the Mechanics of Economic Development', *Journal of Monetary Economics*, vol. 22, no. 1 (March 1988), pp. 3–42.

Mill, J. S., *Principles of Political Economy* (first published 1848; London: Longmans, Green and Company, 1940).

Nelson, R. R., 'Research on Productivity Growth and Productivity Differences: Dead ends or new departures?', *Journal of Economic Literature*, vol. 19, no. 3 (June 1981), pp. 1029–64.

Operations Evaluation Department, *World Bank Support for Industrialization in Korea, India and Indonesia* (Washington, DC: World Bank, 1992).

Pack, H. and L. E. Westphal, 'Industrial Strategy and Technological Change: Theory versus reality', *Journal of Development Economics*, vol. 22, no. 1 (September 1986), pp. 87–128.

Rosenberg, N., *Perspectives on Technology* (Cambridge: Cambridge University Press, 1986).

Streeten, P. P., 'Markets and States: Against minimalism', *World Development*, vol. 21, no. 8 (August 1993), pp. 1281–98.

Taylor, L., 'Review of the World Bank, *World Development Report 1991: The Challenge of Development*', *Economic Development and Cultural Change*, vol. 41, no. 2 (December 1993), pp. 430–41.

Wade, R., *Governing the Market: Economic Theory and the Role of Government in East Asian Industrialization* (Princeton: Princeton University Press, 1990).

Westphal, L. E., 'Industrial Policy in an Export-Propelled Economy: Lessons from South Korea's experience', *Journal of Economic Perspectives*, vol. 4, no. 3 (March 1990), pp. 41–59.

World Bank, *The East Asian Miracle: Economic Growth and Public Policy* (New York: Oxford University Press, 1993).

Young, A., 'Learning by Doing and the Dynamic Effects of International Trade', *Quarterly Journal of Economics*, vol. 56, no. 2 (May 1991), pp. 369–405.

5 Structural Adjustment and African Industry[1]

INTRODUCTION

Over the years, many analysts have despaired of the effects of structural adjustment on the economies of sub-Saharan Africa: the ability of these economies to respond positively to exposure to market liberalisation has been weak and faltering. The weakness of the African supply response has been particularly marked in manufacturing industry and manufactured export performance. Even the World Bank, the main architect of adjustment programmes, has at times shown disappointment with the failure of its recipes, though it has remained doggedly optimistic. However, its latest report on adjustment in Africa (World Bank, 1994) comes to rather optimistic conclusions on the impact of adjustment, especially on industry. Thus, *Adjustment in Africa* starts as follows, 'In the African countries that have undertaken and sustained major policy reform, adjustment is working.'[2] It finds that the countries that had the most improvements in policies had the strongest increases in GDP growth and that the 'increase in their industrial and export growth rates was even more striking'. In view of the past debate on the effects of structural adjustment in Africa[3] (following the established convention, the discussion of Africa covers sub-Saharan Africa excluding South Africa) these findings are important and deserve scrutiny.

There is another, perhaps more important, reason to look closely at this publication. While several analysts have noted some of the drawbacks to the underlying neoliberal view of adjustment,[4] the case of *industrial* response to liberalisation needs further exploration. This chapter places the analysis of adjustment and African industry in the context of the debate on the role of industrial policy in promoting industrialization in the developing world. The Bank's recent book on Africa is especially apt for this purpose, since it is the second in a new series of policy studies addressing the issue of the role of governments in industrial development.[5] This series is intended to present to the world the Bank's considered position on major development issues, and a

124

common theme underlies it: the state should withdraw from economic life, apart from furnishing the rules of the game and 'market-friendly' interventions (to manage the macro-economy well and invest in necessary infrastructure and education). This is the essence of the neoliberal philosophy that characterises the Bank's approach to adjustment. In view of the importance attached to the series and the enormous publicity backing it, it is necessary to look at its findings and methodology.

This chapter examines the analytical underpinnings of structural adjustment and the claim that structural adjustment is benefiting industrial development in Africa. The second section examines the theoretical basis of structural adjustment as formulated and implemented by the World Bank. The fourth section deals with findings of the Bank's study and uses new data to see how industry and manufacturing exports have responded. Since broad comparisons of adjusting countries are difficult to interpret, it goes on to examine the experience of Ghana, the country with the longest history of consistent adjustment in Africa. The final section concludes with some policy considerations. (See Chapter 7 for an expanded treatment of many of these points.)

THE ANALYTICAL BASIS OF ADJUSTMENT AND INDUSTRIALISATION

There is no strict, universally accepted definition of 'structural adjustment'. What is often referred to as structural adjustment really has two different components. One is macroeconomic reform or 'stabilisation', policy changes to achieve internal and external balances in the short to medium term, and generally considered the province of the IMF. The other is 'adjustment' proper, or what Toye (1995) calls 'structural adjustment in the narrow sense': *reforms to free up market forces and so promote long-term growth* ('getting prices right'), the province of the World Bank. SAPs (structural adjustment programmes) can take many forms, but share a common set of premises. In the words of the World Bank, they involve

> unleashing markets so that competition can help improve the allocation of resources ... getting price signals right and creating a climate that allows businesses to respond to those signals in ways that increase the returns to investment.[6]

It is important to distinguish between *stabilisation* and *structural adjustment*, even if the dividing line between the two is blurred in practice (the two Washington institutions share some areas of policy advice). Macroeconomic stabilisation generally precedes or accompanies adjustment, and many stabilisation measures can also constitute important features of adjustment. Exchange-rate adjustment, for instance, can be an important stabilisation as well as adjustment measure, and can have important resource-allocation effects. However, it is generally agreed that adjustment as practised by the World Bank is a set of longer-term policy changes than would be involved in correcting macroeconomic imbalances.[7] For this reason, it is important to look separately at adjustment 'in the narrow sense', to assess the long-term effects of reforming the structure of the economy.

There is another reason. There is little debate in the development literature about the need for good macroeconomic management – stabilisation is almost universally accepted as a policy goal. No government argues that poor macro-policies and instability are conducive to growth, even though governments often find it difficult to achieve the desired balances and differ with the IMF over the form and speed of stabilisation. By contrast, the premises of adjustment *are* the subject of considerable debate. Stabilisation requires that governments act prudently and live within their means, adjustment requires that they withdraw as much as possible from intervening in resource allocation. This chapter takes the need for stabilisation as given, in order to focus on the principles and impact of adjustment.

The principles underlying structural adjustment are derived from simple neoclassical models whose precepts are generally taken to be so self-evident that they are rarely spelled out explicitly. It is difficult, therefore, to find a clear statement in Bank publications of what economic liberalisation is supposed to do in particular circumstances. However, there is a great deal of analysis by the World Bank of the role of industrial policy and selective interventions in promoting industrial development, and the case for reform rests on precisely the same set of arguments, applied to restructuring existing industry rather than to setting up new industry. We can therefore infer from this what SAPs are supposed to do in liberalising economies.

The World Bank's approach is that markets are essentially efficient and that government interventions in resource allocation are essentially distorting and inefficient. The only exceptions allowed are for market failures in the provision of infrastructure and education, where it recommends functional or 'market-friendly' interventions that do not discriminate between activities.[8] Selective interventions, on the other hand, are taken to be 'market-unfriendly' and to distort efficient resource allocation. In the few instances that the case for selective intervention is admitted in theory (as in the *Miracle* study), it is argued that in practice governments cannot intervene efficiently, and that market failures are invariably less costly than government failures. Since the debate on these issues is well known, it is not necessary to go into it at great length here.[9]

Adjustment is based on a number of assumptions which *ensure* that free markets are efficient, leading automatically to an optimal (static) allocation of resources; there is a further assumption that static optimisation in turn leads to dynamic long-term growth. The standard neoclassical conditions are well known – small firms, all 'price takers', operating in competitive markets with no scale economies, perfect information, no risk and uncertainty, and instantaneous adjustment in all markets – and need not be analysed here.[10] In addition, however, there are assumptions on *technology* that are less prominent but are critical to this analysis: these need to be spelt out because they are central to the debate on industrial policy and adjustment.

In this world, all firms operate with full knowledge of all possible technologies, equal access to these technologies and the ability to use technologies efficiently without risk, cost or further effort. The import of technology is just like the purchase of a good: there is a given market price based on perfect information about the product and its competitors, and technology is sold and bought like a physical good. There are no 'tacit' elements in the transfer, no learning costs and no need to make adaptations. Thus, all firms can immediately use technologies with the same degree of efficiency (and all at 'best-practice' levels). In this setting, technical inefficiency is, *ex hypothesi*, due only to managerial slack or incompetence, and can only exist if governments intervene to create barriers to trade or competition.

Where some learning effects in absorbing new technologies are admitted, a simplified view is taken of the learning process. The learning curve is believed to be fairly short and predictable. It is confined to running in a new plant until it reaches rated capacity, and until the benefits of passive learning by repetitive activity are realised. It is generally assumed that such costs are relatively trivial and similar across industries. The learning process does not involve investment, risk or long maturation periods. Firms know what to do to reach best-practice levels: they have full information on what to do to become efficient. Given perfect capital markets, firms can borrow to finance the entire learning process wherever resource endowments justify the technology. If there are capital market failures, these should be tackled at source rather than by governments 'picking winners' and protecting or subsidising them.

Firms are assumed to use technologies as individual units, essentially in isolation. There are no linkages between them, and no externalities resulting from individual efforts to generate skills and information. Since all markets function efficiently, there is no need to create institutions to provide information, technology, and so on. Since there are no externalities, there is no need to coordinate investment decisions across activities that may have intense linkages. Nor are some sets of activities more significant for industrial development than others in that they have more beneficial externalities.

While most development practitioners, including most in the World Bank, would admit that this neoclassical world is not realistic, the design of SAPs appears to take its basic assumptions as *relevant and justifiable for practical purposes*. Thus, it is a premise of adjustment that the immediate and sweeping liberalisation is the most efficient means of economic reform. No activity that is efficient will by definition die out, and none that is inefficient will survive. The demise of inefficient activities will release productive resources for others that are efficient and that will spring up in response to the new price signals. Resources will move with little lag and with no constraints from missing or defective markets. Comparative advantage in industry, as given by resource endowments, will thus be fully realised, not only in a static sense but also in the emergence of new advantages that arise from the accumulation of capital. There is no difference in response between countries apart from that given by different factor endowments, and so no need to design different adjustment programmes according to differing levels of development.

This 'ideal' form of adjustment is rarely found in practice. SAPs differ greatly over time and between countries in their design, content and implementation, depending on the bargaining power and conviction of the governments concerned, and on exogenous events and political expediency.[11] As a consequence, it is difficult to evaluate empirically the effects of SAPs, especially on a specific sector like industry. There are, in addition, numerous problems in separating the impacts of adjustment from those of other factors. In particular, many analysts tend to include stabilisation as part of adjustment – this is suited to certain purposes but does not reveal the impact of long-term measures like liberalisation and the reduction of selective interventions on industrial performance.

These problems are acknowledged but generally ignored in attempts to evaluate the success of SAPs, which tend to proceed by comparing selected indicators of performance of adjusting and non-adjusting economies, or for adjusting countries 'before and after' SAPs. At best some qualifications are made for the degree of adjustment and for some exogenous shocks, but the inherent problem of trying to isolate the impact of a few policy measures (by constructing realistic counterfactuals) remains.[12] It is not the intent in this chapter to discuss such methodological problems. What is attempted here is to narrow the scope of the analysis by clarifying the *underlying theory* of adjustment to see how realistic its assumptions are, and then to trace the impact of those components of the policy package that relate specifically to industrial adjustment (rather than to macroeconomic stabilisation).

A DIFFERENT APPROACH TO INDUSTRIALISATION

Considerable recent research into technological capabilities in developing countries suggests that the neoclassical model that underpins adjustment is oversimplified and misleading.[13] To summarise the main findings of the technological capability (TC) approach: the process of becoming efficient in industry is slow, risky, costly and often prolonged, and faces a range of market failures that may call for interventions in both factor and product markets. In product markets, it may call for infant-industry protection. In factor markets, it may call for interventions to direct resources to particular selected activities, selective as well as functional interventions in skill creation, the promotion of local technological activity rather than a passive dependence on imported technologies, and the setting up of a variety of supporting institutions. In the presence of widespread market failures, simply leaving matters to the market can penalise the development process. In particular, it can hold back entry into activities with complex technologies, increasing the local content and the undertaking of demanding technological tasks locally, what may be broadly labelled 'industrial deepening'.

At the same time, interventions can be distortionary and costly if they are not designed to address market failures, or if they actually create or exacerbate such failures. Import-substituting regimes have been generally marked by such haphazard, pervasive and uneconomic interventions, generally leading to inefficiency and waste rather than to dynamic competitiveness. The TC approach suggests that such patterns of intervention need to be reformed; excessive and distorting interventions should be removed and replaced by policies that address the specific market failures that exist. The larger East Asian NIEs show that selective interventions in an export-oriented setting, carried out by well-trained technocrats and backed by investments in human capital, can be extremely effective in creating a dynamic set of competitive industries, with considerable domestic linkages and indigenous technological content.[14] Without such interventions, industrial development is likely to end up with less dynamism and depth.

The TC approach is as relevant to structural adjustment as it is to the building up of new industries. Inefficient interventions clearly need reform, and an appropriate form of 'structural adjustment' – well designed and truly selective interventions – is needed to replace haphazard import-substituting strategies. However, adjustment based on rapid and sweeping exposure to free markets can kill off activities that are potentially competitive. Such activities also often face learning (or 'relearning') needs which may be risky, costly and prolonged, and which face widespread market failures. If these market failures are not addressed by remedial interventions, the result can be the costly loss of past investments in physical and human capital (in the technological learning that has already taken place).

World Bank-style adjustment can, in other words, remove inefficient government interventions, but *may simply replace government failures by market failures*. If market failures are widespread and remedial interventions are disallowed, the results of adjustment are bound to be disappointing. There is likely to be a weak response in the efficient restructuring of existing activities, in the entry of new activities that can compete in world markets, and, in particular, in the emergence of dynamic new manufactured exports, the biggest dividend that governments hope for from adjusting. In much of sub-Saharan Africa, with severe shortages of skills, especially technical skills relevant to modern industry, where past 'learning' processes have been particularly curtailed by widespread public ownership of industry and the weak tradition of industrial entrepreneurship, the chances of poor results are that much greater.

SAPs AND INDUSTRIALISATION IN AFRICA

Industrial performance has been disappointing in Africa. The growth of manufacturing value added (MVA) from 1980 to 1993 was only

Table 5.1   World Bank findings on MVA growth in adjusting African
countries
(median growth rates for groups)

Country Groups	1981–86	1987–91	Difference
Large improvements	−0.3	4.4	5.8
Small improvements	4.2	5.6	1.2
Deterioration	5.0	5.8	1.1
All countries	3.0	5.5	1.9

Source: World Bank (1994), table A.21.

3 per cent *per annum* in real terms,[15] and the rate of growth declined
steadily over time, from 3.7 per cent in the first half of the 1980s to 2
per cent in 1989–93. The performance was worse in the first part of
the 1990s: MVA growth fell from 3.3 per cent in 1989–90 to 0.4 per
cent in 1991–2, with only a modest recovery to 1.5 per cent in 1992–3.
Moreover, these figures conceal stagnation or falls in MVA in a large
number of countries: many African countries suffered sustained 'deindustrialisation' over the past decade and a half.

The World Bank findings

The World Bank's analysis is based on a sample of twenty-nine
countries in sub-Saharan Africa that have undergone adjustment.
Since many of these adjusting countries have not implemented fully
the reforms recommended, the study divides them into three groups:
six with 'large improvements' in macroeconomic policies, nine with
'small improvement' and eleven with a 'deterioration' (three are un-
classified). The impact of SAPs is assessed by comparing changes in
growth performance in various sectors with reference to these categories, for a pre- and post-adjustment period (1981–6 and 1987–91 respectively). Within groups of countries, the effect of SAPs is shown
by comparing the pre- and post-adjustment periods, while across them
the effect is shown by the extent to which countries improved their
policies (Table 5.1).

These figures suggest that, on average, countries with the largest
improvements in policies (that is, the most adjustment) enjoyed the
largest improvement in manufacturing performance, and those with the
least, the least. Thus, *SAPs are good for industrial growth*. No similar
calculation is done for manufactured exports; the only calculations of
export data relate to total exports.

While these findings appear to show that SAPs benefit African industry, the following points need to be made about the calculation and the conclusions:

- First, the groupings according to improvement or deterioration in policy have *little or nothing to do with adjustment 'in the narrow sense'*. They are based entirely on macroeconomic policy and not on longer-term adjustment processes. If they show anything, it is that improving internal and external balances may be conducive to industrial growth. The impact on resource allocation in response to market orientation cannot be assessed.

- However, the data presented by the Bank do not prove even the beneficial effects of stabilisation. Medians for groups have no statistical significance if individual variations within the groups is larger than the variations between groups. A simple T-test for significance would have resolved the deficiency, but is not provided by the World Bank. Such a test was made on the Bank data for differences between each of the three groups as well as for all improving countries together *vis à vis* deteriorating countries. The test showed that *none* of the groups has significant differences in growth rates in either period, nor are there significant differences between periods. The only exception was the deteriorating group, which had a significantly higher growth rate in the first than in the second period (though the median shows a higher rate of growth in the second period).

- The deteriorating-policy group has higher median growth rates in *both* periods than the other groups (though the difference is not statistically significant), and the large-improvement group has the lowest growth rate in the latter period of adjustment. The lower improvement of the deteriorating group could thus simply be due to the fact that it already had a higher rate of growth. It may also simply reflect the fact that improving countries received higher inflows of scarce foreign exchange in the form of adjustment support, and so were able to improve their performance more, an effect of SAPs that is certainly not part of its publicised benefits.

Trends using updated data

Various aspects of industrial performance in Africa were recalculated using more recent data from the African Development Bank as well as from the World Bank and other sources. The coverage was extended to 45 countries in sub-Saharan Africa, including 16 that had not undergone adjustment (see the full list in the appendix to this chapter). The grouping of adjusting countries by improved or deteriorating policies was, in the absence of other indicators of adjustment, based on the World Bank study. All the countries with 'improved' policies were grouped together, since the differences between the groups with large and small improvements was not found to be statistically meaningful. The periods used were also slightly different. A longer overall period, 1980–93, was taken, with the later sub-period taken to be 1990–93 to capture the more recent effects of policy reforms. The findings of this analysis are as follows.

Rates of GDP growth

The highest rate of growth in both periods is for the adjusting countries with improved macro-policies.[16] The lowest is for the countries with policy deterioration for 1980–93 and for the non-adjusting sub-Saharan African countries during 1990–93. This again suggests that macroeconomic policy improvement seems to help economic growth. However, there is *no significant statistical difference* between any of the group growth rates. Thus the differences in group performance are likely to be caused by other factors than the macroeconomic policy differences that differentiate them.

Manufacturing growth

The growth of manufacturing value added (MVA) shows slightly different trends from that of GDP, and here the relative performance of the policy-improvement group of countries is even better in the post-adjustment period. Table 5.2 shows the data with the trends in the two periods for weighted growth rates. The non-improving groups have higher MVA growth in the period as a whole than in 1990–93, but the policy-deteriorating countries suffer an actual decline in manufacturing in the latter period while the non-adjusters maintain positive rates. By contrast, the policy-improving countries raise their MVA growth rates in the 1990s. A T-test shows that the MVA growth of the policy-improving countries *is* significantly higher than in the policy-deteriorating countries in 1990–93. None of the other differences is statistically significant.

There is now apparently a statistically sound reason to concur with the World Bank's finding that industry in countries with policy improvements performs better than in those with policy deterioration. However, *this proves nothing about the impact of long-term adjustment,*

Table 5.2   MVA growth rates recalculated

Countries	1980–93	1990–93
Policy Improvement	2.7	4.41
Policy Deterioration	1.64	−1.67
Non-adjusting	2.12	0.05

since the groupings only reflect differences in macroeconomic management. And the statistical significance does not establish causation, since a number of important possible influences on manufacturing growth are not taken into consideration. The methodological problems in isolating the effects of adjustment have not been tackled at all.

To check whether it is really liberalisation that is associated with good performance, let us look at the *identity* of the main policy-improving countries with high MVA growth. Of the fifteen countries in this group, only five had annual average increases in MVA of around 4 per cent or more: Burundi (5.7), Kenya (15.5), Mauritania (6.8), Nigeria (4.6) and Uganda (8.1). Of these, Nigeria and Kenya dominate the whole group of policy-improving countries, accounting for 75 per cent of total MVA of this group. *Neither of these countries is considered by the World Bank to have implemented import liberalisation properly.* Both have reversed their trade reforms while their industrial sectors still have significant tariff protection. If they are excluded from the group total, the rate of MVA growth in 1990–93 falls to around 0.1 per cent *per annum*, which equals virtual stagnation – it is certainly not very different from the performance of the non-adjusting countries. Thus, there are scant grounds for arguing that liberalisation led to higher manufacturing growth.

Manufactured exports

The total value of manufactured exports in 1992 by the sub-Saharan countries was only $2.2 billion. Of this total, the policy-improving countries accounted for $1.2 billion, the policy-deteriorating countries for $1 billion and the non-adjusters for only $46 million. The rates of growth for these exports for 1980–92 and 1988–92 are shown in Table 5.3; the period covered is different from previous calculations because of data availability.[17] The table shows that both the improving and deteriorating groups of adjusting countries improved their manufactured-export performance over the period. The best performers for the period as a whole were, however, the policy-*deteriorating* group of

Table 5.3   Growth of manufactured exports

Countries	1980–92	1988–92
Policy Improvement	2.75	9.35
Policy Deterioration	4.66	10.03
Non-adjusters	−8.25	−15.77

countries, not the ones with better macro-management. The non-adjusting countries performed much worse than the others, and their performance worsened over time. T-tests show that the only significant differences are between the non-adjusting countries and the other two groups of adjusting countries in the latter period (but not for the period as a whole).

Can we infer from this that adjustment in the broader sense, including stabilisation, helped manufactured exporters in sub-Saharan Africa? It would appear not, since the best performers are the countries that had worse macroeconomic policies than before. Since the policy-deteriorating countries maintained higher export growth rates in the whole period, this seems to suggest that other factors than stabilisation were at work in affecting export performance. The available data do not, however, allow us to analyse what these were. More importantly for present purposes, the data as they exist do not allow us to even guess at the impact of import liberalisation on manufactured-export growth.

Wages and productivity

Industrial wage and productivity (value added per employee) are shown in Table 5.4 (the Bank study did not touch on these aspects of performance). The non-adjusting countries continued to perform poorly in both respects, with productivity and wages declining over 1980–90. The *best performer is the deteriorating policy group*, with the highest levels and rates of increase in wages and productivity. Statistically there is no significant difference between the groups, but the quality of the adjustment does not seem to raise industrial productivity or wage levels. If anything, the impact of adjustment is negative. As with other such comparisons, however, it is likely that other factors not captured in the data are at work.

Conclusion

This analysis of the data suggests that the World Bank study is unduly sanguine about the effects of policy reform on industry in Africa. In

Table 5.4   Productivity and wages
(levels and annual growth rates)

Country	Average Productivity			Average Annual Wage		
	Value added/Employee ($)		Growth	Wage Level ($)		Growth
	1980	1990	1980–90	1980	1990	1980–90
Policy Improvement	7482	9196	2.32%	2155	2533	1.81%
Policy Deterioration	11374	17321	4.78%	4524	6666	4.40%
Non-adjusters	5461	5300	–0.33%	3093	1925	–5.13%

Source: UNIDO (1994).

fact, nothing can yet be concluded about the effects of adjustment on GDP or industrial growth, export performance or competitiveness. Even the effect of stabilisation on industrial performance is, to say the least, ambiguous: this is not to deny that good macroeconomic policies are desirable in Africa, but to suggest that causal inferences must be drawn with much more care. Clearly there are many missing factors at work and it is not at all apparent what effect adjustment may have had on those. A better way to analyse the impact of structural adjustment may then be to look in detail at the experience of a particular country that has undergone substantial reform to its trade and industrial regime, and to focus on the impact of this form of liberalisation. Let us look at the most relevant one.

The case of Ghana

Ghana has the longest history of consistent adjustment in sub-Saharan Africa, though other countries had earlier adjustment programmes which were not fully implemented. In the World Bank's assessment, Ghana is now the most advanced country in Africa in terms of reaching low tariff-based protection and free trade.[18] The reforms undertaken are impressive: a massive depreciation in the exchange rate (from 2.75 cedis to the dollar in 1982 to 920 cedis in early 1994); the removal of all quantitative restrictions on imports and the lowering of tariffs to a relatively uniform 10–25 per cent range (only luxury products are now at the high end of this range); a reduction of corporate taxes (to 35 per cent) and in capital gains tax (to 5 per cent); the removal of price controls and subsidies; the abolition of credit ceilings and guidelines; the privatisation of state-owned enterprises; the revision of the foreign-investment code; and the granting of incentives to exporters and investors in infrastructure.[19] By the start of the 1990s Ghana had a relatively stable, open and liberal economy, and had achieved *both* reasonable stabilisation and adjustment. It should provide an ideal case for assessing the impact of adjustment on industrial and export performance.

Ghana started its policy reform with an Economic Recovery Programme in 1983. In the initial stages, as far as manufacturing was concerned, this involved freeing up the allocation of foreign exchange for intermediate inputs and spares to an industrial sector starved of access to imports and suffering massive unutilised capacities. There was no direct import competition to Ghanaian industry at this stage. The first SAP started in 1986, and was followed by two others by 1991. It was with these that the process of liberalisation and market orientation was launched. There was a substantial increase in net inflows from foreign sources (mostly in the form of aid), from $196 million in 1985 to an average of $878 million *per annum* over 1989–92.[20] This allowed the economy to finance the newly liberalised imports and to revive domestic demand.

What was the response of the industrial sector? The Bank study shows that the *average* growth rate of manufacturing, negative in the early 1980s, rose to 4.5 per cent per annum over 1987–91 – a healthy response. However, these average data are misleading. MVA did rise rapidly after 1983, when imported inputs were made available to existing industries that were suffering substantial excess capacity. The rate of growth was 12.9 per cent in 1984, 24.3 per cent in 1985, 11.0 per cent in 1986, and 10.0 per cent in 1987. However, as liberalisation spread to other imports and excess capacity was used up, the exposure to world competition led to a steady deceleration of industrial growth. Thus, the rate of growth of MVA fell to 5.1 per cent in 1988, 5.6 per cent in 1989, 1.1 per cent in 1990, 2.6 per cent in 1991 and 1.1 per cent in 1992. These growth figures, in the shape of an inverted-U with a long taper at the end, do not suggest that Ghanaian manufacturing has reached a stage of dynamic takeoff.

It is useful to look at the industrial sector in more detail. Employment in manufacturing fell from a peak of 78,700 in 1987 to 28,000 in 1993.[21] There was a rise in the number of small enterprises, but this was concentrated in low-productivity activities aimed at local markets sheltered from international competition. Foreign investment did not respond to the adjustment, and there was no increase in annual inflows after the SAPs. Moreover, the little that came concentrated on primary activities rather than on manufacturing. Domestic private investment did not pick up sufficiently to lead to a surge of manufacturing growth.

At the same time, large swathes of the manufacturing sector were devastated by import competition. The long period of import-substituting industrialisation in Ghana, with the lead taken by state-owned enterprises, had left a legacy of inefficiency and technological backwardness. It had also left some technological capabilities, but not at such a level that rapid liberalisation could stimulate them to reach world levels in a short period and with relatively low investment. The adverse impact of liberalisation was strongest in the more modern, large-scale part of the industrial sector, which had the most complex technologies and so suffered most from the lack of technological capabilities. Most industrial survivors and new entrants are in activities that have 'natural' protection from imports: very small-scale enterprises, making low-income or localised products, and larger enterprises protected by high transport costs or based on processing of local raw materials.[22]

As far as manufactured exports are concerned, the expectation was that they would grow and diversify rapidly under the new incentive regime and absorb resources released from inefficient import-substituting activities. The data show that while manufactured exports *have* grown since 1986, the values are extremely small, coming to a total of $14.7 million in 1991. There was little sign of a broad-based response on the part of Ghanaian manufacturing enterprises, particularly in its main potential area of comparative advantage, cheap labour. There was practically no diversification of manufactured exports: the growth came mainly from wood and aluminium products, both long-established export sectors, and from firms already established in export markets, rather than from new products or producers.[23]

Labour-intensive exports such as garments, footwear, toys or other light consumer goods, that led the initial export thrust of the Asian NIEs, were conspicuous by their absence in Ghana. Such low-technology 'entry-level' activities, where Ghana should be developing a competitive edge, have been unable to survive the import threat. Conventional wisdom suggests that cheap labour should be the main source of comparative advantage in manufacturing for newly industrialising countries. What this ignores is that even the ability to compete internationally in labour-intensive industries requires a level of productivity and managerial and technical skills that is presently lacking in Ghana. The few relatively well managed firms that exist are largely foreign-owned; among local enterprises the better ones have entrepreneurs that are well educated. The typical local firm, on the other hand, has entrepreneurs with low education, a poorly skilled workforce and no methods for raising their technological capabilities. Most lack the ability even to perceive and define their technological problems.[24]

The theory underlying adjustment does not offer a satisfactory explanation for this phenomenon, since it ignores capability development and the need to overcome market failures in this process. A recent study of technological capabilities in Ghana in the adjustment period concludes that the generally low level of the capabilities has meant that rapid liberalisation, unaccompanied by supply-side measures to develop skills, capabilities and technical support, led to significant and costly deindustrialisation.[25] The growth of new activities and micro-enterprises that is taking place is insufficient to provide a large momentum to growth of production or exports. The expectation that liberalisation by itself will transform Ghana into a 'Tiger' along the lines of East Asia appears facile and unfounded.

Full exposure to market forces in these conditions may actually be retarding the development of Ghana's comparative advantage. The rapid pace of liberalisation is killing off not just inherently uneconomic activities but also some that could be the basis of new manufactured exports. The lack of policies to upgrade skills, technical information and technological support is exacerbating market failures in inputs that are essential for developing competitive capabilities. Ghana's comparative advantage is likely, in this policy framework, to evolve very slowly unless there is a rapid inflow of foreign manufacturing investments. However, the lack of industrial capabilities itself means that foreign investors are not attracted to set up facilities that are export-oriented or are immediately exposed to direct import competition in the domestic market. The only exceptions are activities that have 'natural' protection in the form of a local resource base or high transport costs.

The experience of Ghana clearly has an important bearing on the general issue of industrial adjustment in sub-Saharan Africa: an initial favourable response of manufacturing to adjustment may not lead to sustained growth and diversification if all SAPs do is 'get prices right'. The existence of pervasive market failures raises the costs of adjustment to import competition and holds back the creation of new manufacturing activities and exports. The design of SAPs, by ignoring capability development, places too much store by free markets that work rapidly and effectively. Is Ghana an economically weak 'outlier' whose experience cannot be generalised? It would appear not. Ghana is by African standards a fairly well-endowed and promising economy: it has a relatively good stock of human capital, and its location, resource base and infrastructure would place it in the top quarter of African economies. It has implemented a difficult programme of stabilisation and liberalisation with admirable consistency. All this suggests that its experience illustrates the response of the industrial sector to adjustment more clearly than any country in the region.

POLICY IMPLICATIONS

Policy reform is clearly necessary in Africa, since industrial performance has been abysmal by the standards of the developing world, and the pattern of interventions to support industrialisation has been evidently haphazard and inefficient.[26] Analysis of the capability-building process suggests that they must be replaced by more outward-looking and competitive trade and industrial policies. At the same time, reform of the incentive regime must be integrated with interventions to improve factor markets. Many of these recommendations are in line with what the World Bank would suggest, and they need not be based on the rather dubious statistical analysis that it has provided. However, there remain controversial issues concerning the pace and content of the reform process where we would part company with the Bank. In the present context, there is particular dispute about the efficiency of factor and product markets and the role of the interventions in remedying market failures.

Incentive reform

Let us start with the *incentive system*. An analysis of capability development suggests that a gradual and controlled process of opening up, accompanied by a *strategy* of industrial restructuring and upgrading, is to be preferred to the rapid and sweeping exposure to market forces envisaged by the 'ideal' SAP. The process of liberalisation should be based on a realistic assessment of which activities are viable in the medium term, with the speed of exposure geared to the learning and 'relearning' needs of various activities. The strategy should be developed in collaboration with the industrial sector, and should be pre-announced so that enterprises have time to adjust.[27] Once it is announced, however, the government must stick to the programme so that there is no chance of backsliding and allowing inefficient performers to survive indefinitely.

In this entire process the government should retain powers to guide resource allocation, but in a clear and transparent manner, and with strict requirements in terms of capability development leading to export growth. Unlike earlier strategies of import substitution where governments offered protection with little discrimination between activities, no time limit and no requirements of international competitiveness, this model of adjustment, based on strategies pursued in East Asia, places strong pressures on industries to invest in building up new capabilities to face the import and export competition within a limited period. It is designed to overcome market failures, not to ignore them. It involves close monitoring of the progress of liberalisation, and it requires that the government is able to address the supply-side needs of industries (see below) along with allowing a phased process of liberalisation. The 'ideal' SAP can, by contrast, be extremely wasteful by dispersing and destroying the capabilities already built up and can retard the future development of new and diverse capabilities in industry by ignoring the market failures that exist in this process.

There is little basis, therefore, for the World Bank's conclusion on Africa that 'In promoting exports, governments should not try to pick "winners". . . . Governments can best help entrepreneurs discover and develop competitive exports *by getting out of the way* – and sometimes by helping them along the way.'[28] Similarly, on import liberalisation, it is difficult to justify the ideal for the Bank of low (10 per cent is recommended) and uniform tariff rates for all manufacturing. The Bank does suggest that this should be achieved over the medium term, but *solely for revenue reasons.*[29] It sees no economic reason to retain protection for infant industries, to graduate the liberalisation process according to the speed of 'relearning' and the development of the relevant factor markets, or to discriminate between activities according to their different technological and restructuring needs. As noted, there is neither theoretical nor empirical support for this approach. The main application of the approach, in Ghana, has already been discussed: the move to low and uniform tariffs here was made in *about five years*, which was too rapid for the industrial sector to adjust properly and which yielded disappointing results in terms of growth and competitiveness.

It cannot be too strongly emphasised that to recommend a more gradual and nuanced strategy of liberalisation is *not* to suggest that African countries simply slow down the adjustment process. What is needed is not to put off adjustment, but *to actively prepare* for it in the grace period provided. A vital factor to bear in mind, however, is the strong risk of *government failure*. Most African governments do not at this time have the capability to mount effective interventions in support of industrialisation. It is evident that the launching pad of any reform must be improvements in government capabilities themselves. While it goes beyond the scope of this chapter to discuss this, evidence suggests that these capabilities can be improved by training, reorganisation of the civil service, better performance incentives and monitoring, and greater insulation from the political process.[30] Without such capability development, even the market-friendly strategies recommended by the World Bank have little chance of success. At the same time, it is not recommended that African governments attempt the kind of detailed and pervasive interventions practised in a country like Korea; this does impose tremendous demands on the government and runs very high risks of hijacking and abuse. However, lower levels of direct and detailed intervention such as those used by Taiwan are possible. The correct form of selectivity that can be managed by particular governments is indeed a subject that deserves close study, but is ignored by donors who insist that all selectivity is undesirable.

Supply-side measures

Skill development

The relationship between the lagging industrial development of sub-Saharan Africa and its weak base of human capital has been noted by analysts.[31] Sub-Saharan Africa has the lowest base in the developing world of educational enrolments at practically all levels of schooling, worker training and higher education.[32] Yet this problem is surprisingly absent in the discussion of reform and liberalisation in the region, and in the design of SAPs it does not seem to figure at all. This is so despite the attention paid to 'capability building' in many pronouncements by multilateral institutions in recent years. Liberalisation does not address any of the skill shortages that may be affecting the efficiency of African industry, yet many existing industries might become competitive if their human resources were improved.

The restructuring of industry based on the creation of appropriate skills strengthens the case for a more gradual opening up to international competition. The improvement of the skill base is necessarily a slow process. While African countries may not need very high levels of technical education at the level that their industries are at now, the current supply of technical manpower is so low that even in existing industries there is an insufficient supply of skills to upgrade to world levels of efficiency. The quality of the training, and its relevance to the needs of industrial technology, also needs to be addressed. It is important to note that a certain amount of capability development has already taken place in the African industrial sector, and this is a valuable resource that should be conserved rather than dissipated by a shock therapy that leads to massive deindustrialisation.

It is also important to note that the case is for selective rather than purely functional intervention in skill markets. Given the extreme shortages of the financial and human resources needed to improve education, it is important to concentrate them in the main activities in which adjusting countries can become· internationally competitive. Taking a leaf again from the East Asian book, the successful economies carefully targeted their skill-development programmes to the critical areas of industry where competitiveness could be improved.

The provision of *training* to employees is another vital area of supply-side support. Training services are weak in much of Africa, and enterprises themselves (apart from the major multinationals) invest little in this activity. Even the multinationals train only to the level needed to achieve basic operational efficiency. The apprenticeship system that is prevalent in Africa is more geared to the transmission of traditional skills at fairly low levels of technological sophistication.[33] Reform would need the improvement or setting up of specialised worker and other training institutes, and policies to stimulate greater in-firm investments in training.

Technology support and extension

The development of the technology infrastructure (quality assurance and metrology, research support, information on sources of technology) and the provision of technical extension services to industry, especially to small and medium enterprises (SMEs), is another crucial supply-side measure for industrial upgrading under adjustment. The need for such services is rising in the world; for instance, the requirements of quality control have increased in the past two decades, and international trade in manufactured products increasingly requires stringent quality management.[34] Liberalisation by itself cannot lead firms to the level of quality they need to survive in world markets. A concerted effort is needed by the government to strengthen standards and metrology institutions, to provide consultancy assistance to firms to obtain certification, and to mount a campaign to persuade firms to invest in this process. Little of this is happening in most of sub-Saharan Africa and certainly SAPs include no provision for it.

Similarly, the technical extension services that exist in much of Africa are largely ineffective in providing the inputs that enterprises need to upgrade to competitive levels, as are means to provide research support for industry. Technological levels are very low in most SMEs, and most firms are not even aware of what their deficiencies are or how to set about remedying them. Large firms conduct practically no in-house research and development in Africa. There is very little interaction in general between the industrial sector and the technology infrastructure that exists in many countries to provide R&D and technical support to enterprises. Many of the research institutes are poorly funded, and so have inadequate equipment and unmotivated staff. They do not go out to search for and offer solutions to the technical problems of industry, preferring a more isolated existence. Much the same happens for technology information services to help local firms, especially SMEs, to locate and purchase foreign technologies.

Yet a strong and proactive technology infrastructure is invaluable for upgrading the competitive capabilities of industry. All the governments of East Asia (including Hong Kong[35]) invest a great deal in providing a range of technological and information services to their enterprises; many of these have to be subsidised because of the public-goods nature of their work or because of the inability of small firms to pay full cost. If adjustment is to be successful in Africa, all such services need to be established and strengthened. Again, in view of resource scarcity, the intervention will have to be partially selective: the main thrust should be to improve services to industry 'clusters' of greatest importance to international competitiveness and growth.

Finance and other support

The need to provide adequate financial support for industrial restructuring and upgrading is obvious. Unfortunately, adjustment in Africa takes place mostly under conditions of extreme financial stringency, exacerbating the effects of rapid liberalisation and of inadequate measures to improve competitiveness. Information services for potential exporters are generally weak. The physical infrastructure is poor and has worsened over time.

Many of these recommendations are probably acceptable to the World Bank. However, the problem is that they are not integrated into adjustment programmes, but undertaken, if at all, in separate operations. Their de-linking from the process of incentive reform may be because the proponents of rapid liberalisation believe that the benefits of exposure to market forces will accrue regardless of whether or not supply response is boosted, or that supply-side improvements should only be undertaken when the market is giving the 'right' signals. However, this chapter has argued that this approach is costly and inefficient, and based on faulty reasoning on the nature of the market failures that confront industrial enterprises trying to become competitive.

To conclude, therefore, there is no controversy on the need for stabilisation and some form of policy reform in industry. The debate is really about how efficient markets are, and what role governments should play in improving them. On the arguments advanced here, market failures are rife and structural adjustment must be pursued more gradually, with greater control, strategy and involvement by the government. Clearly, many African governments presently lack the capability to mount effective strategies. However, given the need to remedy market deficiencies to achieve sustainable industrial development, the *first step in adjustment should be to improve their intervention capabilities*. The evidence on the benefits of past SAPs on industrial growth and diversification does not provide much support for the present World Bank 'line'. It is therefore regrettable that the world's premier development institution considers it necessary to pursue such a line. Perhaps the design of adjustment should start with a serious reconsideration of its own conceptual apparatus and beliefs.

APPENDIX: AFRICAN COUNTRIES BY POLICY ON STRUCTURAL
ADJUSTMENT

Improvements in Policy	Deterioration in Policy	Non-Adjusters
Burkina Faso	Benin	Angola
Burundi	Cameroon	Botswana
Gambia	CAR	Cape Verde
Ghana	Chad	Comoros
Kenya	Congo	Djibouti
Madagascar	Côte D'Ivoire	Equ. Guinea
Malawi	Gabon	Ethiopia
Mali	Guinea	Lesotho
Mauritania	Guinea Bissau	Liberia
Niger	Mozambique	Namibia
Nigeria	Rwanda	São Tomé
Senegal	Sierra Leone	Seychelles
Tanzania	Togo	Somalia
Uganda	Zambia	Sudan
Zimbabwe		Swaziland
		Zaire

REFERENCES

African Development Bank, *African Development Report 1994* (Abidjan: African Development Bank, 1994).
De Valk, Peter, 'A Review of Research Literature on Industry in Sub-Saharan Africa under Structural Adjustment', in R. van der Hoeven and F. van der Kraaij (eds), *Structural Adjustment and Beyond in Sub-Saharan Africa* (London: Curry, 1994), pp. 227–39.
Elbadawi, I. A., 'Have World Bank-supported Adjustment Programmes improved Economic Performance in Sub-Saharan Africa?' (Washington, DC: World Bank, 1992), Policy Research, Working papers, WPS 1001.
Elbadawi, I. A., Ghura, D. and Uwujaren, G., 'Why Structural Adjustment has not Succeeded in Sub-Saharan Africa' (Washington, DC: World Bank, 1992), Policy Research, Working papers, WPS 1000.
Helleiner, Gerald K., 'The IMF, the World Bank, and Africa's Adjustment and External Debt Problems: An unofficial view', *World Development*, vol. 20, no. 6 (1992), pp. 779–92.
Helleiner, Gerald K., 'From Adjustment to Development in sub-Saharan Africa', *UNCTAD Review* (1994), pp. 143–54.
Husain, Ishrat, 'Structural Adjustment and the Long-term Development of Sub-Saharan Africa', in R. van der Hoeven and F. van der Kraaij (eds), *Structural Adjustment and Beyond in Sub-Saharan Africa* (London: Curry, 1994), pp. 150–71.
Husain, Ishrat and Faruqee, Rashid (eds), *Adjustment in Africa: Lessons from the Country Studies* (Washington, DC: World Bank, 1994).
Killick, Tony, *The Adaptive Economy: Adjustment Policies in Small, Low-Income Countries* (Washington, DC: World Bank, EDI Development Series, 1993).
Lall, Sanjaya, 'Structural Problems of African Industry', in F. Stewart, S. Lall and S. Wangwe (eds), *Alternative Development Strategies in Sub-Saharan Africa* (London: Macmillan, 1992.a), pp. 103–44.
Lall, Sanjaya, 'Technological Capabilities and Industrialization', *World Development*, vol. 20, no. 2 (1992.b), pp. 165–86.
Lall, Sanjaya, 'Trade Policies for Development: A policy prescription for Africa', *Development Policy Review*, vol. 11, no. 1 (1993), pp. 7–65.
Lall, Sanjaya, 'Industrial Policy: The role of government in promoting industrial and technological development', *UNCTAD Review* (1994), pp. 65–89.
Lall, Sanjaya, Navaretti, Giorgio Barba, Teitel, Simon and Wignaraja, Ganeshan, *Technology and Enterprise Development: Ghana under Structural Adjustment* (London: Macmillan, 1994).
Leechor, C. 'Ghana: Forerunner in adjustment', in Husain and Faruqee, *Adjustment in Africa* (1994), pp. 153–93.
Mosley, P., Harrigan, J. and Toye, J., *Aid and Power: The World Bank and Policy-based Lending* (London: Routledge, 1991).
Mosley, Paul and Weeks, John, 'Has Recovery Begun? "Africa's adjustment in the 1980s" revisited', *World Development*, vol. 21, no. 10 (1993), pp. 1583–606.
Mosley, Paul and Weeks, John, 'Adjustment in Africa', *Development Policy Review*, vol. 12, no. 3 (1994), pp. 319–27.

Singh, Ajit, 'Openness and the Market Friendly Approach to Development: Learning the right lessons from development experience', *World Development*, vol. 22, no. 12 (1994), pp. 1811–23.

Stein, Howard, 'The World Bank and the Application of Asian Industrial Policy to Africa: Theoretical considerations', *Journal of International Development*, vol. 5, no. 1 (1994.a), pp. 1–19.

Stein, Howard, 'Theories of Institutions and Economic Reform in Africa', *World Development*, vol. 22, no. 12 (1994.b), pp. 1833–50.

Stein, Howard, 'Deindustrialization, Adjustment, the World Bank and the IMF in Africa', *World Development*, vol. 20, no. 1 (1992), pp. 83–95.

Toye, John, *Structural Adjustment and Employment Policy: Issues and Experience* (Geneva: International Labour Office, 1995).

UNIDO, *Industry and Development: Global Report 1993/94* (Vienna: United Nations Industrial Development Organisation, 1994).

World Bank, *Sub-Saharan Africa: From Crisis to Sustainable Growth* (Washington, DC: World Bank, 1989).

World Bank, *World Development Report 1991* (Washington, DC: World Bank, 1991).

World Bank, *The East Asian Miracle: Economic Growth and Public Policy* (New York: Oxford University Press, 1993).

World Bank, *Adjustment in Africa: Reforms, Results, and the Road Ahead* (New York: Oxford University Press, 1994).

6 Malaysia: Industrial Success and the Role of Government

BACKGROUND ON INDUSTRIAL PERFORMANCE

The Malaysian economy has been one of the best performers in the developing world over the past twenty-five years. Malaysian GDP grew at an annual rate of 6.7 per cent during 1971–90, led by a manufacturing sector that expanded at 10.3 per cent. Performance was even stronger in the early part of the 1990s, when the economy grew at 8.1 per cent *per annum* and the manufacturing sector at 12.3 per cent, the highest rate of industrial growth in Asia with the exception of China. This industrial performance led to a massive structural transformation of the economy, with the share of manufacturing in GDP rising from 14 per cent in 1971 to 30 per cent in 1993, and that of traditional sectors (agriculture and mining) declining from 43 to 24 per cent.[1] This impressive performance has two quite distinct segments – the export of manufactured products and production for the domestic market.

Manufactured export growth

The early source of export earnings for Malaysia was processed natural resources like rubber, tin and palm oil; the facilities were initially largely foreign-owned but came under local ownership over time. The growth of non-traditional manufactured exports started in the 1970s, with the entry of electronics firms from the USA. Manufactured exports then expanded rapidly, growing at a compound 24 per cent *per annum* during 1971–90. There was a setback during 1981–6, when the rate of growth slowed down (if only to 16 per cent *per annum*), but it was followed by a recovery (to 29 per cent) during 1987–92. The share of manufactures in total exports rose from 12 per cent in 1970 to 71 per cent in 1993, when they reached a total of $34 billion. This performance took Malaysia from being a predominantly resource-based exporter to being the developing world's sixth largest exporter of manufactures, just behind the Four Dragons of East Asia and China.

More interestingly, Malaysian manufactured exports developed on the basis of what are regarded as high-skill and technologically complex products. While processed natural resources provided the early impetus for industrial exports, electrical and electronic products accounted for the bulk of its export growth after the 1970s. This was unlike the typical low-technology exports like garments, footwear and toys that drove the early growth of most other NIEs. This base was expanded and diversified over time, and by the 1990s electrical and electronic products accounted for nearly 60 per cent of its manufactured exports. As a result, Malaysia emerged as the world's largest exporter of semiconductors, and among the largest exporters of disk drives, telecommunications apparatus, audio equipment, room air-conditioners, calculators, colour televisions and various household electrical appliances. A simple classification of the skill-intensity of exports based on wages in the relevant industries in the USA[2] (above average wages counting as high-skill and below average wages as low-skill) shows that 73 per cent of its manufactured exports in 1980, and 84 per cent in 1990, were in the high-skill category.[3] In terms of values, Malaysian high-skill exports increased from $1.6 billion in 1980 to $13 billion in 1990.

This performance sets Malaysia off from its neighbouring 'new NIEs': Thailand had high-skill exports of $0.7 billion in 1980 (45 per cent of its total manufactured exports) and $3.6 billion in 1990 (48 per cent). Indonesia lagged much further behind, with $0.2 billion (45 per cent) in 1980 and $1.4 billion (a large decline in share to 16 per cent of a rapidly growing export base in garments, footwear and plywood) in 1990. Malaysia's high-skill ratio in 1990 was even higher than Korea's (65 per cent) though the absolute value of the latter's high-skill exports ($30 billion) was much larger. The Malaysian pattern of export growth followed most closely that of neighbouring Singapore, which also made a rapid transition to high-skill products quite early in its industrialisation process and then maintained its momentum by strategic interventions to ensure further upgrading into high-technology products.

However, such classifications by the factor content of exports can be somewhat misleading. Malaysia's initial entry into electronics in the 1970s was in fact highly labour-intensive, based on the manual assembly of semiconductors. It was followed after some time by similar assembly operations in audio and other electronic and electrical products. Low wages and a favourable investment climate thus accounted for Malaysia's early export growth, as they did in the other new NIEs. As wages rose in Malaysia and technologies advanced abroad, however, export operations were not moved to cheaper countries. They were automated and more complex processes and products were introduced. Thus, highly capital-intensive processes replaced manual semiconductor assembly, and more sophisticated tasks like semiconductor testing were shifted from the developed countries. Other electronics firms, particularly from Japan and Taiwan, transferred entire production lines for mature products to Malaysia (like the assembly of colour TVs, room air-conditioners, calculators, and a wide range of audio equipment). New entrants came with complex technologies (such as hard disk drives for computers). Many foreign component suppliers, again mainly from East Asia, followed their principals and invested in Malaysia in the export-processing enclaves.

Over time, therefore, Malaysia's export structure did become more skill-intensive. It did not, however, attain the levels of local research, design or linkages of countries like Korea, Singapore or Taiwan. Even today, Malaysia remains at the assembly stage of many electrical and electronics products. Many of the consumer products based there are the relatively mature stage of their product cycles. The export sector remains relatively isolated from the rest of the economy and lacks the technological base that has developed in other NIEs. In comparison with the other 'new NIEs', however, Malaysia upgraded very successfully, and there are exceptions where import-substituting industries are maturing into international competitiveness, particularly in the public sector.

Domestic-market-oriented industrialisation

While exports led Malaysian industrialisation in the 1980s, the domestic market was the main driving force in industrial growth in the two previous decades. In common with other developing countries, Malaysia pursued an import-substitution policy to initiate its industrialisation process, and even after launching its export drive it maintained a substantial protected industrial sector to serve domestic markets. The domestic market, driven by primary exports, accounted for 40 per cent of the increase in industrial growth in 1959–68 and for 144 per cent of the increase in 1973–81.[4] Import substitution accounted for 51 per cent in the former period and for 36 per cent in the second, while export growth accounted for 10 per cent and *minus* 80 per cent respectively. In the 1980s the roles were reversed, and exports accounted for a 138 per cent growth while the contributions of both domestic demand and import substitution turned negative. However, the base of a significant part of Malaysian industry was laid in the previous period, and was expanded in the 1980s by a new bout of import substitution in heavy industry led by the public sector.[5]

Import-substituting in Malaysia had something of a dual structure. In the domestic private sector, the initial period of import substitution was dominated by light (first-stage) assembly and packaging activity by local Chinese-owned firms. There was little deepening of this sector into second-stage import substitution, for several reasons: the domestic market was relatively small, the level of local technological capabilities low, and levels of protection moderate by developing-country standards (in the late 1970s the average effective rate of protection for industry was 31 per cent and declined to 17 per cent by 1987[6]). Indigenous Malay-owned industry was largely small-scale and traditional, concentrated in the rural areas. The ethnic tensions that existed between the Malay government and the Chinese business sector may also have held back concerted efforts to promote the local private sector in more complex industries. Much of the policy effort in the private sector was to encourage the growth of Malay entrepreneurship and employment.

The main sources of domestic-oriented heavy industry in Malaysia were state-owned and multinational facilities in a number of intermediate and capital-goods industries. Despite protection and other forms of encouragement, the growth of heavy industry had not met expectations in the first decade or so of independence. The drive was consequently strengthened in 1980, when the government set up the Heavy Industries Corporation of Malaysia (HICOM). The objective of HICOM was to diversify manufacturing activity (and reduce the heavy dependence on a small number of export-oriented activities), create more local linkages (which the export-oriented enclaves were failing to do), promote small and medium enterprises (particularly those owned by Malays), and lead technological development by collaborating with foreign firms and investing in local R&D. The concept was based on the industrial policies pursued in countries like Korea and Japan, though these countries, unlike Malaysia, relied upon the private sector to spearhead the drive.

The experience of Malaysian state-owned industry was mixed. In general, financial returns were poor or negative and the lack of experience and capabilities led to prolonged teething problems. The deteriorating macroeconomic situation in the mid-1980s induced the government to launch a gradual process of privatisation and restructuring of state-owned enterprises. However, the most strategic parts of HICOM, like automobile manufacturing, were kept under state ownership, but were restructured to improve management and make it more market-oriented. The rates of return improved in later years, with better management and improvements in local suppliers, macroeconomic recovery and entry into export markets. An analysis of the trade statistics shows that some of the import-substituting industries enjoyed high rates of export growth in the 1980s: during 1980–90 exports by the chemical industry grew by 20 per cent *per annum*, those by non-metallic minerals by 26 per cent, and by iron and steel (1985–90) by 33 per cent. In 1990 the total value of these exports came to around $1 billion. Automobile exports started much more recently, and in 1991 reached $153 million; since then they have grown vigorously. While these still constitute a small part of Malaysia's total manufactured exports, it is encouraging to note that some infant industries are 'growing up' – this point is taken up later.

THE ISSUE OF INDUSTRIAL POLICY

What lessons does the Malaysian case have for industrial policy? Clearly its rates of growth of manufacturing output and exports, and their concentration in high-technology activities, are very impressive and, if replicable, could be a model for other industrialising countries. What can they be traced to? Were they caused by market forces in which the government played no particular role? Were they the results of government intervention, and could they have been even better without intervention? Or was there an element ('luck') that neither interventions nor liberalisation can account for, and so cannot be reproduced elsewhere?

These issues have been brought to the fore in the recent debates on industrial policy, particularly in East Asia.[7] The 'pure' neoclassical (neoliberal) school holds that governments played a minimal role in the rapid growth enjoyed by the NIEs, that markets were essentially efficient and were left to work their magic by governments that furnished macro-stability and basic infrastructure. A more moderate ('market-friendly') neoclassical view holds that all markets were not efficient, and that governments played a more positive role – by intervening to create human capital and by strongly pushing exports. This modified neoliberal view of government is espoused by the World Bank in its analysis of the East Asian 'miracle', and is now the official stance of major aid donors.[8]

A distinction is drawn in this literature between 'functional' interventions, where only generic failures in factor markets are remedied, and 'selective' interventions, where particular activities are chosen for encouragement over others. Admittedly, both forms were prevalent in

Asia, but with large differences between countries in their choice of form. The interventions that accounted for the East Asian success were, in the Bank's view, overwhelmingly functional. The use of selective interventions (with the relatively minor exception of capital-market interventions) rarely helped development, and governments got away with such policies only because their bureaucracies were insulated from the political system, well trained and motivated. But other governments lack these conditions, and so should not contemplate selectivity – corruptibility is inherent to all other governments and raises the costs of industrial policy to unacceptable levels. The *Miracle* study concludes that the real lesson of East Asia is that governments should, apart from creating human capital, follow sound macroeconomic policies and pursue export orientation. If market failures exist, they are always less costly than government failures: there is never a case for selectivity.

The World Bank's approach to the case for industrial policy and its interpretation of the evidence in the *Miracle* study have been strongly criticised.[9] There are a number of gaps in the discussion; evidence is selectively used; the econometric analysis is flawed; the relevant theory is not fully developed, and even its rudimentary analysis of market failures is not related to the facts; and too much is made of the uniqueness of the Asian political economy. In particular, it fails to take into account the basic nature of the industrial learning process in developing countries – this process is highly variable by technology and can be costly, risky and prolonged, and may confront a variety of market failures unless it is undertaken by foreign firms that can internalise most of these failures.[10] The deepening of industrial structure and the development of advanced technological capabilities and linkages thus require that supportive, and selective, interventions be undertaken. These forms of deepening progressed very differently in East Asia, in response to differing interventions. It is therefore these *differences* that are directly relevant to the debate on industrial policy. Yet the *Miracle* study deliberately ignores them in order to minimise the potential benefits of industrial policy.[11] It therefore smacks of ideology rather than reasoned analysis. Here is not the place to enter this broader debate – what is relevant to the present discussion is the nature and terminology of the discussion.

MALAYSIAN INDUSTRIAL POLICY

With independence in 1957, Malaysia inherited a liberal economic system with a well-functioning bureaucracy, a tradition of stable macroeconomic policies, good infrastructure, reasonable living standards and a developed resource-based sector that provided most of its export earnings.[12] The economy was strongly trade-oriented and foreign investors played a large role in the productive sectors. The new government kept most of these initial conditions intact, in particular a well managed macroeconomy, a welcoming attitude to foreign investors and openness to trade. However, it also launched various measures to stimulate economic and industrial growth.

Analysts trace three phases of Malaysian development.

(a) In the first phase, 1957–70, a moderate degree of protection was given for import-substituting activities, and measures to attract FDI into export activity were launched and strengthened. While the main thrust of policies was to promote rural development and infrastructure (Salleh and Meyanathan, 1993), industrial interventions grew over the period. In the area of export promotion, the setting up of the Malaysian Industrial Development Authority (MIDA) in the late 1960s was an important step. It was at this time that the semiconductor-assembly boom in developing countries was reaching its peak, and Singapore was trying to upgrade from labour-intensive assembly to more complex activities. MIDA spotted the opportunity and made a concerted effort to target electronics MNCs in the USA. In the absence of a previous record of such activity in Malaysia, it was largely the targeting of MIDA, coupled with generous fiscal incentives, the disciplined, literate and English-speaking labour force and the favourable investment climate, that launched Malaysia on its high-tech export path.[13]

On the *domestic* front, import substitution policies accounted for much of the growth in industrial production in this period. The bulk of industry was Chinese-owned and was largely concentrated in light, first-stage manufacturing. By the end of this period, the small domestic market was showing signs of saturation and industry was failing to create rapid employment growth. There were few linkages between the booming export sector and domestic enterprises, and few domestic-oriented manufacturing activities were showing signs of graduating independently to international competitiveness.

(b) The second phase, 1971–85, was marked by the launch of the New Economic Policy (NEP). The NEP focused on improving the economic situation of the *bumiputeras* (indigenous Malays) in the wake of racial disturbances in 1969. While the earlier policy of moderate import substitution continued, NEP involved several new domestic industrial interventions. Among the most important was the taking of increasing domestic shares in foreign-owned plantations and non-export enterprises and the setting up of state enterprises, to be later transferred to Malay private ownership, to foster local supplier industries and create new industrial skills. By the end of this period there were some 700 state-owned enterprises in a variety of economic activities. These enterprises (as well as Malay-owned small enterprises in general) were given strong preferences in government procurement and financing. Employment and education quotas were used to improve the economic base of the Malay population, and in 1975 the passing of the Investment Coordination Act strengthened measures to raise *bumiputeras'* participation.

The continuing shallowness of the MNC-led export sector and its lack of linkages with the rest of industry (especially Malay enterprises) caused the Mahathir government to launch stronger industrial-policy measures in the early 1980s. The main tool was the heavy-industry drive led by HICOM (noted above). Malaysian policies in this period 'looked East', and followed the HCI (Heavy and Chemical Industry) drive of Korea in the 1970s. 'The HICOM projects were intended to push Malaysia into diversification, create modern manufacturing activity outside FTZ enclaves, and foster linkages between industries. State involvement was necessary to overcome private investors' caution about high-cost, high-risk ventures with long gestation periods.'[14]

HICOM suffered large losses in the early years of its operation; in addition many of the other smaller state-owned enterprises also performed poorly. HICOM's investments were in highly demanding activities, and it is not surprising that the small indigenous base of domestic technological and managerial capabilities (that held back private entry into heavy industry) led to inefficiencies and a long 'teething period' for many of its enterprises. As noted, the government modified its strategy after macroeconomic difficulties in the latter half of the 1980s. Many observers (such as the World Bank, 1993) regard this foray into heavy industry as a costly failure, and take the government's scaling down of the effort as a repudiation of Malaysian industrial policy in general. This is unwarranted, for two reasons: the design of the interventions in Malaysia was not ideal, and so does not constitute a proper test of the effectiveness of industrial policy; and the period over which effectiveness should be assessed may need to be longer when complex learning processes are involved.

It is evident that the objectives and modalities of Malaysian industrial interventions were quite different from those of the Korean government in the HCI drive.[15] First, the NEP was addressed primarily to redressing social imbalances and not to gaining world-market competitiveness in a new set of industrial activities. HICOM and other state industrial enterprises were set up to serve domestic markets and establish local linkages, and there was no systematic attempt to guide and monitor their technological development process. In contrast, Korean heavy industry was largely set up to create new bases of export competitiveness, and its progress was disciplined and monitored. Protected enterprises were forced to enter world markets at a relatively early stage, though in complex technologies the process of maturing sometimes took a decade or more.[16]

Secondly, the Malaysian interventions were led by public-sector enterprises with 'soft budgets', without a previous 'learning' base for absorbing complex organisational and production technologies. In Korea the drive was led by the giant private conglomerates (the *chaebol*), which already had a strong and diverse production and export base, and had to perform in competitive markets. The difference between the two need hardly be stressed in the context of developing competitive capabilities. However, as the Malaysian enterprises accumulated such capabilities and were restructured to become more market-oriented, their performance improved. The experience of East Asia suggests that the results of entry into heavy industry should be assessed over a fairly long period, and in this sense the 'jury is still out' on whether Malaysian interventions have been successful.

Thirdly, the Malaysian interventions were not backed up by supply-side measures to ensure adequate skill development or technology support. While possessing a good base of basic education, Malaysia had a relatively small base of technical education at the university and vocational levels: today one of the most critical constraints to its industrial upgrading lies in the small base of high-level technical and engineering skills.[17] Moreover, it had a strong institutional base to support plantation and mining activities, but lacked an effective support system for industrial-technology development and extension. This meant not only that state-owned firms had weak capabilities, but also that the linkages that they were supposed to create took longer to establish. These also contrasted strongly with Korea, where supply-side measures to support its industrial deepening proceeded hand in hand with interventions in trade and investment. As noted below, however, the Malaysian government is now launching similar supporting measures to promote competitive advantages.

On the export front, FDI in export-oriented manufacturing continued to be encouraged during this phase, and Malaysia's favourable image attracted increasing numbers of investors from the USA and Japan, and to a lesser extent from Europe and the Asian NIEs. However, as noted, the export sector remained relatively isolated from the rest of the economy and highly concentrated in a relatively small number of electrical and electronic products. This concentration led to a sharp decline in exports in the mid-1980s in response to a recession in the developed countries – the effect on Malaysia was much greater than in East Asia where the NIEs had a more diverse base of manufactured exports. During this period, there was a significant upgrading of process technology towards automation as wages rose and technologies evolved. However, the lack of a strong technical skill base in Malaysia, and the reluctance of MNCs to transfer design and research activities to developing countries, meant that operations continued to be based largely on final-assembly operations. Again, a strong contrast may be drawn to the larger East Asian NIEs, where much of the export drive came from domestic enterprises and the strong skill base coupled with interventions to promote local R&D and procurement led to far stronger technological development and linkages.

(c) The last phase, from 1985 onwards, started with the recession noted above. This induced the Malaysian government to adopt more pragmatic strategies towards racial balance and to rein in the public sector. It started a drive to privatise and restructure public enterprises and launched a fresh drive to attract MNCs. The Promotion of Investments Act of 1986 gave fresh, more generous incentives for manufacturing and other investments, and some of the ethnic requirements of the NEP were relaxed. FDI responded vigorously, and in the latter half of the 1980s showed the rising importance of the new NICs, led by Taiwan and Hong Kong, as foreign investors. Japanese MNCs continued to relocate their assembly operation in Malaysia as the Yen strengthened, and induced many of their suppliers to invest along with them.

On industrial policy, the government replaced NEP with the New Development Policy (NDP), which moved *much closer to the kind of industrial interventions practised by the East Asian NIEs*. Industrial policy was now based on a more coherent and systematic analysis of the needs and capabilities of manufacturing activities. An Industrial Master Plan was drawn up (the first IMP ran for 1986–95, and a new IMP is being formulated for the next decade) to target activities for strengthening comparative advantages. The emphasis shifted towards a more selective strategy, i.e. providing critical factors for industrial development (skills and training, technical support, finance, quality improvement[18]) and more targeted import protection. A technology action plan was drawn up to strengthen the infrastructure of science and technology institutions and stimulate R&D in private enterprises. Industrial restructuring programmes were devised to provide cheap finance for the wood, textiles and engineering industries. Trade was liberalised over time (by 1994 most activities were offered tariff protection of 20 per cent), but infant-industry protection was retained, and continues to be used: one recent example is the launching of light-aircraft production in the public sector. The restructured public sector retained a role in industries (such as the automotive, petrochemical, iron and steel and cement industries) where the required investments are large and long gestation periods are involved.

At the same time, the government also moved to more selective policies on export-oriented MNCs. MIDA started to use incentives to guide FDI into higher value-added activities and more technology-intensive processes, very much in the Singaporean mould. Prospective investors in areas of advanced technology were targeted, not just by MIDA but also by state development corporations (the Penang Development Corporation, in the heart of Malaysia's electronics region, is particularly active). A 'Growth Triangle' was set up with Singapore and Indonesia to attract new investments into Johor, and another triangle is planned in the north (with Thailand and the Sumatra province of Indonesia) to promote outward investment by labour-intensive activities from Penang. In addition, unlike Singapore, the Malaysians introduced additional incentives to increase local content (foreign suppliers that had invested in Malaysian export-processing zones to supply their principals were to be denied the full privileges they used to receive as wholly export-oriented firms, and were to be treated as local firms).

A recent study by the Asian Development Bank summarises the new approach as follows:

> Under the IMP, industrial policy instruments have been consolidated and now focus more narrowly on attempts to promote reinvestments, industrial linkages, exports and training. Moreover, the IMP allows for a continuous review of the promoted sectors and orchestrates efforts at industrial rationalisation and restructuring where there is evidence of declining competitiveness. A similar approach was adopted by Korea in the reform of its HCI drive in the early 1980s. The corporate tax structure now has fewer exemptions, allowing for a more effective use of selective incentives. . . . In recent years, the government has moved to ensure that trade policy is designed more

in line with overall development strategy. Protection has been governed by dynamic notions of comparative advantage, promoting the development of industrial subsectors that are intended to replace light manufacturing activity as the main exporters.[19]

The new strategy has been accompanied by strong export and income growth, though much of it is due to the response of MNCs to strong incentives and the rise in the value of the Yen and in the costs of production in the NIEs. Some of the import-substituting heavy industries are starting to export, and there are encouraging signs of the 'maturing' of infant industries as technological and managerial capabilities develop. It is too early to judge whether they can be a dynamic long-term source of growth of exports, but certainly at this time they are offering higher local value added and linkages than the traditional export enclaves.

EVALUATING MALAYSIAN INDUSTRIAL POLICY

Malaysian industrial policy is taken in the World Bank's *Miracle* study to have been largely a failure. Industrial policy is identified in the study with the state-led, heavy industry drive in the 1980s. The failure of this policy is taken to be established by the fact that the government reversed its policy on state ownership after the high costs of the drive became evident.[20] The Bank attributes Malaysian (and East Asian) industrial success to conformance to free-market forces in an export-oriented, stable macroeconomic setting, and to functional interventions in infrastructure and human capital formation. There is an element of truth in this, as the discussion of its export performance suggested. However, the neoliberal argument can be overdone.

The deficiencies of this argument in the case of the Dragons and Japan have been considered elsewhere. As far as Malaysia goes, the argument here is that stability and export orientation, combined with a basic set of skills, were certainly important in Malaysian success. However, there were other elements that have to be considered before a general conclusion on the inefficacy and redundancy of selective interventions is drawn.

First, the early growth of manufacturing industry in Malaysia was fostered by import substitution and provided much of the impetus for industrial development up to the 1970s. The benefits of this era of industrial policy have to be taken into account. The later phase of industrial policy, the state-led entry into heavy industry, did have higher costs. However, the subsequent reduction of the role of the Malaysian state in industry, precipitated by the macroeconomic crisis of the mid-1980s, does not establish the failure of industrial policy *per se*. All it shows is that this particular form of industrial policy in the 1980s may not have been well designed. As noted above, the Malaysian heavy industry drive lacked many of the critical elements of East Asian policies in terms of overcoming the market failures that confront industrial deepening. The political economy of Malaysia at the time dictated a different, and apparently less effective, set of interventions. But direct state ownership is not really necessary for industrial policy, since interventions can be just as pervasive without such ownership. The reduction of this ownership and the imposition of market discipline, which the Bank sees as a renunciation of interventions, should be seen instead as an improvement of industrial policy.

Secondly, the experience of Japan and Korea suggests that the period over which the success of heavy-industry policy is to be judged should be longer than allowed in the Malaysian case. The development of a complex activity like automobile manufacturing can take one to two decades if there is to be significant local content and design, even in countries with long industrial traditions (as in Latin America). In a country with a relatively shallow industrial base, like Malaysia, it is creditable that Proton has done as well as it has in its relatively short life (it started in 1983). It has increased local content to over 66 per cent, when other assemblers are still at around 30 per cent; in the process it has fostered a large number of local suppliers by an active technology-transfer programme. It has mounted a successful export drive. It has introduced new models with increasing local design content, and created new engineering capabilities in the country. Its long-term success remains to be established, but it is certainly too early to write it off as a costly failure. There are a number of other heavy industries that have also established an export capability and show signs of 'growing up' from infancy.

Thirdly, functional interventions and a stable macro-framework were certainly central to Malaysia's success in attracting export-oriented FDI. However, it must be noted that selective targeting by MIDA was important in getting the electronics-based boom started. In addition, there was a strong element of luck – Malaysia's legacy of good physical and legal infrastructure, its location next to Singapore and its entry into electronics at the height of the semiconductor-assembly boom. These benefits are not available to other developing countries that are liberalising today and seek to attract export-oriented FDI. The pattern of investment is thus very different, and Malaysia increasingly looks unique in its leap into high-technology exports.

Fourthly, Malaysia's export success has certain peculiar features that worry policy makers.[21] In brief, these are:

- First, a very high degree of *concentration*, in terms of reliance on a few manufactured products to drive exports. In 1990 the top five products accounted for 58.9 per cent of total exports (at the three-digit SITC level), as compared, say, with 39.6 per cent for Korea and 45.4 per cent for Japan.

- Second, *MNCs* continue to dominate Malaysian exports, and on a rough estimate provide over three-quarters of the total value of manufactured exports.[22] There is relatively low direct participation in export activity by local, especially private, firms, the main exception in the private sector being garments (garments now account for around 6 per cent of manufactured exports).

- Third, the *local content* of most manufactured exports remains low. Apart from local resource-based products, most exporters have relatively weak linkages with the domestic economy. The import content has increased from 35 per cent in 1970 to about 42 per cent in 1990. Where there has been increase in local content over time, it has been largely driven by foreign investment in the component sectors.

- Fourth, local enterprises, whether domestic or foreign, perform relatively few high value-added and technologically demanding tasks such as design or development. After many years of production, most exporters remain heavily dependent on foreign sources of technology, not just for major innovations (which is the case with all developing, and many industrialised, economies) but also for minor improvements, equipment, design and specialised skills. The bulk of export activity in Malaysia is in relatively simple assembly and finishing operations, even though some more advanced activities like testing have taken root and many of the facilities are highly automated. The general level of Malaysian technological capabilities is low relative to the NIEs, and the underlying supply of high-level technical skills is inadequate.

- Fifth, Malaysian exporters have not developed *independent marketing capabilities*, constraining their ability to upgrade into higher value-added products and markets. Foreign affiliates rely on their parent companies' marketing networks and brand names. Local firms in textiles and garments sell mainly through foreign buyers, and a substantial portion of Malaysian exports are routed through traders in Singapore.

The concern of the government is essentially that the concentrated, MNC-dominated export structure, with low technological and supply linkages to the local economy, provides an unstable basis for export growth, and may not be sufficient to drive the structural upgrading the government desires.[23] MNCs have not so far responded to rising wage and land costs by moving abroad. On the contrary, they have invested in creating a base of operational skills and physical facilities in Malaysia that would be expensive to reproduce elsewhere. However, the danger is less that they would relocate their existing facilities than that over the longer term their *new* investments in higher-technology activities would concentrate on other locations. Many developing countries with lower wage costs are now actively seeking foreign investments, and some, like China and India, can offer huge domestic markets as well as a larger base of engineering skills, domestic suppliers and service firms. Some countries in Eastern Europe may become attractive sites for high-technology investments in the medium term.

Though FDI inflows are highly variable, recent trends of FDI in high-technology activities coming into Malaysia are not encouraging.[24] Even if they can be revived, the structural problems noted above will remain. The experience of NIEs like Taiwan and Singapore suggests that active selective interventions will be needed to upgrade FDI and to make it strike deeper technological roots and linkages in Malaysia. Many of these resemble the interventions launched in recent years by the Malaysian government, but it remains to be seen if the capabilities can be developed to achieve results comparable to those of these other countries.

Fifthly, following from the previous point, many factor-market interventions needed to maintain industrial growth in Malaysia are selective rather than functional, and should not be counted as 'market-friendly'. Thus, the creation of specific skills being launched in Malaysia for electronics design, biotechnology or automobile design is a selective process. So is the setting up of technical-support services for textiles or food processing, or the encouragement of R&D in the strategic technologies identified in the technology plan. Such selectivity in factor-market interventions constitutes improved industrial policy in Malaysia, not a renunciation of such policy.

Sixthly, the Malaysian government is setting up the mechanisms for the kinds of private–public cooperation and interaction that were so effective in East Asia.[25] The launching of the concept of 'Malaysia Incorporated', with a forum for regular consultation between the government and private business, marks an important step forward in mounting effective design, monitoring and implementation of interventions.

Finally, it is also relevant to note that Malaysia is reaping the benefits of industrial policy undertaken earlier by *other* countries in the region, in particular Japan, Taiwan, Singapore and Korea.[26] A large part of its export growth is fuelled by the capabilities that selective interventions in these countries fostered, which then spilled over to neighbouring countries that offered cheaper labour and a conducive investment climate, or, as with Singapore, from policies to encourage MNCs to move their lower value-added facilities overseas. The market forces that Malaysia's 'market-friendly' policies tapped were thus themselves the creatures of industrial intervention. While this is not of direct concern to Malaysia (except as models on which it draws), it *is* relevant to the debate on industrial policy more broadly.

To conclude, therefore, the evidence suggests that while much of Malaysia's growth in recent years was driven by FDI and market-friendly policies, neoliberal conclusions – that selective interventions were irrelevant or positively harmful in Malaysia – seem unwarranted. Industrial policy in Malaysia did play a role, and has an important role to play in the future, though its pattern has differed greatly from the old NIEs. Import substitution in early years laid the basis of manufacturing industry in the country. Targeting was involved in getting the export boom started, and is being used to further upgrade the export sector. The heavy-industry drive should have been designed and implemented differently, but the need for a concerted set of interventions to promote industrial diversification was valid. The need for selective as well as functional interventions to strengthen the local skill and technological base is correctly, and strongly, recognised by the government. The change in direction of the government in recent years, combining liberalisation of trade with more targeted selective interventions in both trade and factor markets in concert with the private sector, shows an improvement in industrial policy rather than its absence. Finally, the benefits that are spilling over to Malaysia from industrial policy in the old NIEs and Japan should be counted as benefits of intervention rather than of free markets.

REFERENCES

Ali, A. (1992), *Malaysia's Industrialization: The Quest for Technology* (Singapore: Oxford University Press).

Amsden, A. (1989), *Asia's Next Giant* (Oxford: Oxford University Press).

Amsden, A. (1994), 'Why Isn't the Whole World Experimenting with the East Asian Model to Develop? Review of *The East Asian Miracle*', *World Development*, 22, pp. 627–34.

Ariff, M. (1991), *The Malaysian Economy: Pacific Connections* (Singapore: Oxford University Press).

Brown, J. (1993), 'The Role of the State in Economic Development: Theory, the East Asian experience, and the Malaysian case' (Manila: Asian Development Bank, Economics Staff Paper No. 52).

Chang, Ha-Joon (1994), *The Political Economy of Industrial Policy* (London: Macmillan).

Colclough, C. and Manor, J. (eds) (1992), *States and Markets: Neo-Liberalism and the Development Policy Debate* (Oxford: Oxford University Press).

Dutt, A. K., Kim, K. S. and Singh, A. (eds) (1994), *The State, Markets and Development: Beyond the Neoclassical Dichotomy* (London: Edward Elgar).

Edwards, C. and Jomo, K. S. (1993), 'Policy Options for Malaysian Industrialisation', in K. S. Jomo (ed.), *Industrialising Malaysia: Policy, Performance and Prospects* (London: Routledge), pp. 316–34.

Government of Malaysia (1991), *The Second Outline Perspective Plan, 1991–2000* (Kuala Lumpur: National Printing Department).

Hobday, M. (1995), *Innovation in East Asia: The Challenge to Japan* (London: Edward Elgar).

Jacobsson, S. (1993), 'The Length of the Learning Period: Evidence from the Korean Engineering Industry', *World Development*, 21, pp. 407–20.

Jomo, K. S. (ed.) (1993), *Industrialising Malaysia: Policy, Performance and Prospects* (London: Routledge).

Kwon, J. (1994), 'The East Asian Challenge to Neoclassical Orthodoxy', *World Development*, 22, pp. 635–44.

Lall, S. (1992), 'Technological Capabilities and Industrialization', *World Development*, 20, pp. 165–86.

Lall, S. (1994.a), 'Industrial Policy: The role of government in promoting industrial and technological development', *UNCTAD Review 1994*, pp. 65–89.

Lall, S. (1994.b,) ''The East Asian Miracle'' Study: Does the Bell Toll for Industrial Strategy?', *World Development*, 22, pp. 645–54.

Lall. S., Hamid, N., Boumphrey, R. and Hitchcock, B. J. (1994), *Malaysia's Export Performance and its Sustainability* (Manila: Asian Development Bank, draft).

MITI (1993), *Malaysia: International Trade and Industry Report* (Kuala Lumpur: Ministry of International Trade and Industry).

Moreira, M. M. (1994), *Industrialization, Trade and Market Failures: The Role of Government Intervention in Brazil and the Republic of Korea* (London: Macmillan).

MOSTE (1990), *Industrial Technology Development: A National Plan of Action* (Kuala Lumpur: Ministry of Science, Technology and the Environment).

Rodrik, D. (1994), 'King Kong Meets Godzilla: The World Bank and *The*

East Asian Miracle', in A. Fishlow *et al.*, *Miracle or Design: Lessons from the East Asian Experience* (Washington, DC: Overseas Development Council).

Salleh, I. M. and Meyanathan, S. D. (1993), 'Malaysia: Growth, Equity and Structural Transformation' (Washington, DC: World Bank), *The Lessons of East Asia*.

UNCTAD (1994), *Trade and Development Report 1994* (Geneva: UNCTAD).

Wade, R. (1990), *Governing the Market* (Princeton: Princeton University Press).

Wade, R. (1994), 'Selective Industrial Policies in East Asia: Is *The East Asian Miracle* right?', in A. Fishlow, C. Gwin, S. Haggard, D. Rodrik and R. Wade, *Miracle or Design? Lessons from the East Asian Experience* (Washington, DC: Overseas Development Council).

Westphal, L. E. (1990), 'Industrial Policy in an Export-Propelled Economy: Lessons from South Korea's experience', *Journal of Economic Perspectives*, 4, pp. 41–59.

World Bank (1993), *The East Asian Miracle: Economic Growth and Public Policy* (New York: Oxford University Press).

7 Skills and Capabilities in Ghana's Competitiveness

(with Ganeshan Wignaraja)

INTRODUCTION

Ghana has shown a weak industrial-supply response to structural adjustment, particularly in terms of manufactured growth and competitiveness. Before adjustment policies were implemented, the reasons offered for the stagnation and decline in the Ghanaian industrial sector were external shocks, political uncertainty, macroeconomic mismanagement, hostility to private (domestic and foreign) firms, a grossly inefficient public sector and an inward-looking trade regime.[1] These analyses called for better governance, stabilisation, import liberalisation, privatisation and more openness to foreign direct investments – the staple of adjustment programmes the world over (see Chapter 5). Most analysts predicted that the adoption of liberal market-friendly policies would *by itself* be enough to revitalise industrial growth and export expansion. There was some expectation, not least by the Ghanaian government, that Ghana would become an East Asian style 'Tiger' by implementing a full-blown structural adjustment programme.

There is little doubt that external shocks and poor policies did have a negative effect on Ghana's industrial performance. However, these did not fully explain the extent of uncompetitiveness and the lack of dynamism in much of Ghanaian manufacturing. Now that adjustment has been in place for nearly a decade, and manufactured export performance remains weak and the base of exports narrow and confined to a few resource-based activities, it is time to look for more fundamental explanations for competitive weaknesses. These explanations are based on an examination of the skills and knowledge needed to set up and efficiently operate modern industry, what is broadly termed 'technological capabilities' (TCs). It is argued that it is the low level of TCs that explains why so much of Ghana's manufacturing remains confined to the lowest end of the technology scale and why it responds so poorly to the incentives that the structural adjustment has provided.

To the extent that this is so, policy remedies have to address other issues than those normally considered by analysts and donors.

This chapter draws on case studies of 32 manufacturing firms in four industries (textiles and garments, metal working, wood working and food processing) to provide a qualitative analysis of TCs in Ghana. It considers the principal influences on technological development, including entrepreneurship, technical manpower and training.[2] Where possible, comparisons are drawn with other developing countries.

APPROACHES TO INDUSTRIAL COMPETITIVENESS

In recent years, there has been increasing recognition among economists that an essential ingredient in the competitive advantage of nations is *the ability of manufacturing firms to become technically efficient in a world of constantly changing technologies.*[3] While there are many theories of comparative advantage that include technology as a major determinant, almost all take the micro-level process by which technology is mastered and improved for granted. Their focus on major technological innovation and frontier technology as sources of comparative advantage ignores the fact that 'minor' technical change may be equally important as a source of competitive advantage. More importantly from the viewpoint of developing countries, it ignores the fact that even to use a given technology efficiently may involve a long, difficult and costly process of learning. This section examines the assumptions on technological capabilities (TCs) that underlie standard theories of comparative advantage. It then introduces the TC approach based on empirical research on firm-level technological development in developing countries.

In the standard neoclassical approach all markets are assumed efficient. Product markets give the correct signals for investment in new activities and factor markets respond to these signals without serious lags or friction. At the firm level, given perfect competition, information, foresight and efficient factor markets, the optimum point on the production-possibility frontier is chosen according to prevailing factor prices. All firms are by definition equally efficient: technology is freely available, with full knowledge on techniques available to all firms – most importantly, it is costlessly and instantly absorbed, and any learning process is known, predictable and automatic. Over time, as factor prices change to reflect changing endowments, the firms' activities change accordingly – this represents the optimal pattern of specialisation and forms the basis for evolving comparative advantage. With these assumptions, it follows logically that interventions can only be distortionary. There is, however, nothing in neoclassical theory which says that if the assumptions are changed and market imperfections admitted, its welfare and policy conclusions remain the same. The extent to which the initial assumptions apply in practice is an empirical question. It is neoclassical development economists who, generally implicitly, have tended to assume that markets in developing countries are in fact efficient, and that imperfections are of little practical or policy significance.

For the technological-capability approach it may be useful to describe briefly the micro-level perspective on industrial development. This provides a more realistic and complete framework for the analysis of market failures and of the need for interventions than the simple neoclassical model based on a unique static equilibrium.[4] Industrial competitiveness in developing countries depends essentially on how well individual firms manage the process of technological and managerial development. Technology is not perfectly transferable like a physical product: it has many 'tacit' elements that need the buyer to invest in developing new skills and technical and organisational information. Technological development thus does not mean innovating new technologies (though this is clearly one end of a broad spectrum of technological effort) but, at least at the start, it means efficiently using imported technologies.

The process of gaining technological competence is not instantaneous, costless or automatic, even if the technology is well diffused elsewhere. It is risky and unpredictable, and often itself has to be learnt: in developing countries new firms may not even know what their deficiencies are or how to go about remedying them. The development of competitive capabilities may be costly and prolonged, depending on the complexity and scale of the technology. It involves interactions with other firms and institutions: apart from physical inputs, it calls for various new skills from the education system and training institutes, technical information and services, contract research facilities, interactions with equipment suppliers and consultants, standards bodies, and so on. The setting up of this dense network of cooperation needs the development of special skills. This constant and uncertain process of learning differs radically from the standard neoclassical model of firm development, and leads to different policy implications.

Industrial development is not just about starting new activities. As economies progress and mature, it involves '*deepening*' in any or all of four forms – technological upgrading of products and processes within industries, entry into more complex and demanding new activities, increasing local content, and mastering more complex technological tasks within industries (from those relevant to assembly to those needed for more value-added activity, adaptation, improvement, and finally design, development and innovation). Each involves its own learning costs. These costs differ by activity, rising with the sophistication of the technology, the extent of linkages and the level of technological capabilities aimed at. Progressive deepening is to some extent a natural part of industrial development, but it is not inevitable. Its pattern and incidence differ greatly, *depending on the strategies pursued by the government.*

Industrial progress in developing countries depends *essentially on how well firms manage this complex process of technological development.* The process of capability development may face various *market failures.* Free markets may not, in other words, give correct signals to resource allocation between simple and difficult activities or between investments in importing technology and internal technological effort. The first is the basis of the classic case for infant-industry protection: *in the presence of learning costs, a latecomer to industry necessarily faces a disadvantage compared with those that have undergone the learning process.* This disadvantage may be exacerbated by the lack of appropriability of some forms of investment in learning (externalities in the form of skills and information that leak out to other firms), the technological linkages with other activities that are also undergoing uncertain learning processes, and the absence of developed capital markets capable of financing the learning process.

Given these costs, cumulative learning effects, externalities, unpredictabilities and capital markets' imperfections, all endemic to developing countries, exposure to full import competition can prevent entry into activities with relatively difficult technologies. Thus, interventions may be necessary to induce the deepening of technologies that may be in the country's longer-term comparative advantage.

However, intervention in the form of protection against competition may itself take away the incentive to invest in learning. Moreover, widespread and open-ended protection to a large range of activities may result in little learning, general inefficiency and a lack of industrial dynamism. Since protection penalises consumers and downstream industries, it is imperative that this cost be offset by dynamic gains in learning and spillovers to make it economically worthwhile to intervene. Efficient industrial policy requires that protection be *limited* in extent and duration, and that its deleterious effects be *offset* by measures to force firms to invest in developing their technological capabilities,

and by containing its effects so that export activities are not handi-
capped.

The most effective way to do this is to combine domestic protection
with strong export orientation, providing a cushion for learning along
with incentives to being fully competitive, and letting export activities
operate in an effective free-trade regime as far as their access to inputs
is concerned. This particular design of protection to infant industries
constitutes the crucial difference between 'classic' import-substituting
regimes, which promoted some learning but distorted its direction and
dynamism, and the aggressively export-oriented regimes of East Asia,
that combined extensive and variable protection with powerful push to
their firms to enter world markets.

The insights yielded by the technological-capability approach to industrial competitiveness can be applied to structural adjustment. Inefficient interventions result in truncated and distorted learning, and reform
is clearly needed in 'classic' import-substituting regimes. However,
rapid exposure to import competition can kill off activities that are
potentially competitive but are not given the time or the resources to
complete their learning process or to 'unlearn' past distorted learning
and become competitive. There is a costly learning process involved
in *adjusting* to competition if past interventions have been excessive
and have not created the environment for healthy learning in industry.
Reforms have therefore to be gradual, based on relearning needs and
guided by an overall strategy.

Since learning costs differ between activities, interventions and liberalisation have to be *selective* rather than uniform. This goes against
the basic neoclassical tenet that protection should never be discriminatory, based on its simplified view of passive and uniform learning
across activities. In simple activities the need for protection may be
minimal, because the learning period is relatively brief, easy to get
information on, and predictable. In complex activities, with large scales,
advanced information and skill needs, wide linkages and intricate organisations, by contrast, the learning process could spread over years,
even decades. These may never be undertaken (unless there is a strong
natural-resource cost advantage) unless protection is given.

Since the factor needs of technology development, as for new skills
and information, differ by activity, interventions in factor markets have
to be *integrated* to interventions to protect or promote activities. Factor markets often fail in developing countries, and such failures are
more readily accepted by institutions like the World Bank. One of the
main concessions to the need for policy is in the field of education,

where the Bank agrees that markets fail but suggests that interventions are 'market-friendly' because they do not discriminate between activities. While this is true of a part of education at the basic levels, interventions in higher education and training can be highly selective if they are geared to the specific needs of industries being targeted for promotion. For instance, a government setting up an electronics industry has to target the training of electronics engineers and technicians – this is exactly what the NIEs of Asia did – if its overall policy is to succeed. The identification of 'market-friendly' interventions with education and technology is untenable, since such interventions can be highly selective.

SAPs AND INDUSTRIAL COMPETITIVENESS IN GHANA

Structural adjustment programmes (SAPs) as designed and implemented by the World Bank are strongly based on the neoclassical approach outlined above. They thus start from the presumption that markets are essentially efficient in developing countries, and government interventions in resource allocation essentially distorting and inefficient. The only exceptions allowed by the World Bank are market failures in the provision of infrastructure and education, where it recommends functional or 'market-friendly' interventions that do not discriminate between activities.[5] Selective interventions, on the other hand, are taken to be 'market-unfriendly' and to distort efficient resource allocation. In most such writings no case is admitted for infant-industry protection or for other sources of market failure that can call for selective interventions to restore market efficiency. In the few instances that such cases are admitted in theory, it is argued that in practice governments cannot intervene efficiently, and that market failures are invariably less costly than government failures. Since the debate on these issues is well known, it is not necessary to go into it at great length here.[6]

The 'ideal' form of structural adjustment recommended by the Bank follows logically and forcefully from these assumptions. No empirical evidence is needed, and indeed none is sought, to support the argument. To simplify, this takes the form of five prescriptions:

- Remove all forms of selective intervention and restore free market-driven resource allocation (often referred to as 'getting prices right'). In the trade arena, expose industrial activities to international competition, as a precondition to other adjustment measures. Allow free entry to foreign private investment, and exercise as little discretion as possible in international investment flows. In the domestic arena, promote liberal entry, exit, ownership, and flexible labour markets; privatise public enterprises wherever possible, restructure them where not.

- Get prices right in all economies in the same manner regardless of the level of development, since by definition all markets are efficient (or more efficient than governments).

- Carry out reforms quickly and across the board, since there is no economic justification for continuing to differentiate between activities. No 'strategy' is needed to guide the restructuring or upgrading process at the level of industry or firms since markets will give the correct signals.

- Do not link the pace of reforms of the incentive structure to measures to improve human capital or infrastructure, since this will take much longer and, in any case, factor markets will also respond better if the overall price signals are correct.

- Finally, having got rid of the legacy of inefficient interventions, do not retain any further scope for selective interventions to promote industrial growth.

This 'ideal' form of adjustment that the World Bank would like to implement is rarely found in practice. SAPs differ greatly over time and between countries in their design, content and implementation, depending on the bargaining power and conviction of the governments concerned, and on exogenous events and political expediency.[7] As a consequence, it is difficult to evaluate empirically the effects of SAPs, especially on a specific sector like industry. There are, in addition, numerous problems in separating the impact of adjustment from those of other factors. In particular, many analysts tend to include stabilisation as part of adjustment – this is suited to certain purposes but cannot reveal the impact of liberalisation on industrial performance.

Many analyses have been carried out, particularly in Africa, to assess the impact of SAPs. They usually involve quantitative comparisons of adjusting economies with non-adjusting ones, or of adjusting economies before and after adjustment (or sometimes a combination).[8] Needless to say, these comparisons are fraught with a number of methodological, data and analytical problems, and it has proved extremely difficult to establish a clear causal relationship between adjustment measures and economic performance while taking account of all other influences on performance.

It may therefore be more instructive to look in greater detail at the experience of Ghana, the country with the longest history of consistent adjustment in sub-Saharan Africa. This analysis is not just a mechanistic comparison of 'before and after' figures, but a micro-level analysis within a clear TC framework that allows us to interpret and explain the efficiency and supply response of the industrial sector.

In the World Bank's assessment, Ghana is now the most advanced country in Africa in terms of adjustment, and has come closest to low tariff-based protection and free trade.[9] The reforms undertaken are impressive: a massive depreciation in the exchange rate (from 2.75 cedis to the dollar in 1982 to 920 cedis in early 1994); the removal of all quantitative restrictions on imports and the lowering of tariffs to a relatively uniform 10–25 per cent range (only luxury products are at the high end of this range); a reduction of corporate taxes (to 35 per cent) and in capital gains tax (to 5 per cent); the removal of price controls and subsidies; the abolition of credit ceilings and guidelines; the privatisation of state-owned enterprises; the revision of the foreign-investment code; and the granting of incentives to exporters and investors in infrastructure.[10] By the start of the 1990s Ghana had a relatively stable, open and liberal economy, and was often referred to as a 'model' adjuster in Africa.

Ghana started its policy reform with an Economic Recovery Programme in 1983. In the initial stages, as far as manufacturing was concerned, this involved freeing up the allocation of foreign exchange for intermediate inputs and spares. There was no direct import competition to Ghanaian industry at this stage. The first World Bank structural adjustment programme started in 1986, and was followed by two others until 1991. It was over these SAPs that the process of liberalisation and market orientation was launched. There was a substantial increase in net inflows from foreign sources (mostly in the form of aid), from $196 million in 1985 to an average of $878 million *per annum* over 1989–92.[11] This allowed the economy to finance imports and to revive domestic demand.

What was the response of the industrial sector? Data in the Bank study show that the average growth rate of manufacturing, negative in the first half of the 1980s, rose to 4.5 per cent *per annum* over 1987–91 – the predicted positive response. However, this average is misleading. MVA did rise rapidly after 1983, when imported inputs were made available to existing industries that were suffering substantial excess capacity. The rate of growth was 12.9 per cent in 1984, 24.3 per cent in 1985, 11.0 per cent in 1986, and 10.0 per cent in 1987. However, as liberalisation spread to other imports and excess capacity was used up, the exposure to world competition led to a steady deceleration of industrial growth. Thus, the rate of growth of MVA fell to 5.1 per cent in 1988, 5.6 per cent in 1989, 1.1 per cent in 1990, 2.6 per cent in 1991 and 1.1 per cent in 1992. These growth figures, in the shape of an inverted-U with a long taper in recent years, do not suggest that Ghanaian manufacturing has reached a stage of dynamic takeoff.

Employment in manufacturing fell from a peak of 78,700 in 1987 to 28,000 in 1993.[12] There was a rise in the number of small enterprises, but this was mainly in low-productivity activities aimed at local markets, sheltered from international competition, and not enough to lead to longer-term growth and competitiveness. Foreign investment did not respond to the adjustment, and there was no increase in annual inflows after the SAPs; the little that came concentrated on primary activities rather than on manufacturing. At the same time, large swathes of the manufacturing sector were devastated by import competition. The long period of import-substituting industrialisation in Ghana, with the lead taken by state-owned enterprises, left a legacy of inefficiency and technological backwardness. It also left some technological capabilities, but not at a level where rapid liberalisation could stimulate them to reach world levels in a short period and with relatively low investment. The adverse impact of liberalisation was strongest in the more modern, large-scale part of the industrial sector, which had the most complex technologies and so suffered most from the lack of technological capabilities. Most industrial survivors and new entrants are in activities that have 'natural' protection from imports: very small-scale enterprises, making low-income or localised products, and larger enterprises protected by high transport costs or based on processing of local raw materials.[13]

As far as manufactured exports are concerned, the expectation was that they would grow and diversify rapidly under the new incentive regime and absorb resources released from inefficient import-substituting activities. The data show that while manufactured exports *have* grown since 1986, the values are extremely small, coming to a total of $14.7 million in 1991. There was little sign of a broad-based response on the part of Ghanaian manufacturing enterprises, particularly in its main potential area of comparative advantage, cheap labour. There was practically no diversification of manufactured exports: the growth came mainly from wood and aluminium products, both long established export sectors, and from firms already established in export markets, rather than from new products or producers.[14]

Labour-intensive exports such as garments, footwear, toys or other light consumer goods, that led the initial export thrust of the Asian NIEs, were conspicuous by their absence in Ghana. Such low-technology 'entry-level' activities, where Ghana should be developing a competitive edge, have been unable to survive the import threat. Conventional wisdom suggests that cheap labour should be the main source of comparative advantage in manufacturing for newly industrialising countries. What this ignores is that even the ability to compete internationally in labour-intensive industries requires a level of productivity and managerial and technical skills that is presently lacking in Ghana. The few relatively well managed firms that exist are largely foreign-owned; among local enterprises the better ones have entrepreneurs that are well educated. The typical local firm, on the other hand, has entrepreneurs with low education, a poorly skilled workforce and no methods for raising their technological capabilities. Most lack the ability even to perceive and define their technological problems.[15] The growth of new activities and micro-enterprises that is taking place is insufficient to provide a large momentum to growth of production or exports. The expectation that liberalisation by itself will transform Ghana into a 'Tiger' along the lines of East Asia appears facile and unfounded.

Exposure to market forces in these conditions may actually be retarding the development of Ghana's comparative advantage. The rapid pace of exposure to world competition is killing off not just inherently uneconomic activities but also some that could be the basis of new manufactured exports. The lack of policies to upgrade skills, technical information and technological support is exacerbating market failures in inputs that are essential for developing competitive capabilities. Ghana's comparative advantage is likely, in this policy framework, to evolve very slowly unless there is a rapid inflow of foreign manufacturing investments. However, the lack of industrial capabilities itself means that foreign investors are not attracted to set up facilities that are immediately exposed to direct import competition.

TECHNOLOGY CASE STUDIES IN GHANA

The analysis of technological capabilities in Ghana draws on the results of a survey of manufacturing firms. The survey, undertaken in 1992, consisted of 32 firms in four industries: textiles and garments, wood working, metal working and food processing. The purpose of the survey was to analyse the process of acquiring TCs in firms, the technological strengths and weaknesses of different types of firms, and the influences on technological-capability acquisition. The study covered a mix of firms from different size classes, ownership forms (foreign, local non-African, local African) and differing performance categories (growing, reviving, stagnant and dead).

The study involved classifying the technological performance of the sample firms into three categories (investment, production and linkage),[16] and making a detailed evaluation of their capabilities in each. This analysis was used to draw up a list of technologically 'competent' firms. The choice was based on a combination of indicators, since much of the data involved were qualitative. Most of the competent firms were easy to identify, by virtue of their clearly superior investment and production-engineering capabilities. The list, impressionistic though it is, is similar to industry analyses of industrial competence widely used by international consultants.

Characteristics of technologically competent firms

Of the 32 firms, 13 (41 per cent of the sample) were identified as relatively competent. It has to be emphasised that inclusion in this list *does not mean that these firms are technologically capable by world standards*. On the contrary, the evidence suggests that the level of technological mastery by Ghanaian firms of the technologies they use is relatively poor. There is little or no process or product development by even the best sample firms that can be regarded as 'innovative'. The kinds of 'minor innovation' that have been found, in many more industrialised developing countries, to lead to the raising of machine productivity beyond its design capacity – the use of completely different raw materials, the development of technologically complex new products and so on – are rarely found in the Ghanaian sample. In general, the best that the competent firms do is to use imported technologies relatively well and make some adaptations to local circumstances. Table 7.1 shows the characteristics of the competent and other firms and reports the results of some statistical-significance tests.

Let us start with the *general features* of the competent firms. Of the 13, the largest number (5) are in the food processing industry, followed by metal working (4); textiles and garments, and wood working have 2 firms each. While one should not read too much into the industrial distribution of such a small sample of firms, it is worth remarking

Table 7.1 Technologically competent and other Ghanaian firms

	Emp. (nos.)	Age in prod. in 1992 (years)	Capacity utilisation rate (%)	% Non-African/ foreign equity	Aver. wage (US$)	Entr. educn (years) (b)	Prod. Man. educn (years)	Scien., eng. & techn. (% of emp.)	Eng. only (% of emp.)	QC/ mainten. manpower (% of emp.)
Technologically Competent Firms										
Observations	13	13	12	13	13	13	13	13	13	11
Mean	192.3	19.0	66.0	35.4	64.6	17.1	15.5	6.7	1.6	6.5
Standard deviation	176.1	10.7	29.3	38.3	21.4	3.2	2.4	6.2	1.7	3.4
Other Firms										
Observations	18	18	13	18	18	18	18	17	18	18
Mean	69.7	22.7	32.0	18.9	43.9	11.7	8.9	2.8	0.9	1.8
Standard deviation	81.0	14.9	16.9	34.5	15.9	3.4	6.5	2.7	2.3	3.5
T-statistic (a)	2.6*	-0.76	3.6*	1.26	3.09*	4.50*	3.47*	2.25*	0.93	3.15*

Notes: (*a*) *denotes statistical significance at the 5% level.
(*b*) The number of years of education were computed as follows: middle school (8 years), secondary school (12 years), diploma (14 years), BSc (17 years), MSc (20 years), PhD (22 years).

Source: Lall *et al.* (1994).

on the fact that the two activities in which Ghana may be expected on *a priori* grounds, and on the basis of the experience of other developing regions, to have a comparative advantage are garments (which is low-technology and labour-intensive) and furniture (which is local resource-based and labour-intensive). Yet these activities seem generally to register low levels of TC, not just in the sample but also in the industry more generally.

The food-processing and metal-working industries, with relatively more competent firms, are essentially oriented to the domestic market. In addition, many of the competent firms in metal working, normally an engine of technological development, are in relatively simple technologies. There is little sign that more advanced engineering activities are emerging in Ghana. The implications of these trends for future growth and export dynamism are not very promising.

The statistical findings are as follows:

1. *Technologically competent firms are larger than other firms.* There is a statistically significant difference between the average *employment size* of the two groups, with the mean for the competent firms coming to 192 employees and for the other firms 70 employees. Of the thirteen firms in the table, eight are large (over 100 employees), three are medium-sized (30–99) and only two are small (below 30). The correlation between size and competence may indicate one or more of three things. First, it may indicate that firms have reached large size *because* they were competent, i.e. they had earlier invested in TC development to a greater extent, or more effectively, than other firms. Secondly, it may be due to the distribution of activities and technologies in the sample, i.e. in many of the technologies covered there were economies of specialisation and size that meant that only large firms could reach efficient levels of TCs. Thirdly, it may reflect the existence of market segmentation, i.e. only firms above a certain size were able to gain access to the skills, information and credit needed to be competent.

It is not possible to say firmly which of these explanations has most validity, and there is probably some validity in each. The distribution of the firms in the case studies may in some cases have led to the association between size and competence, particularly (given the scale-intensive processes used in modern food processing) in the sample food-processing firms. Even in these technologies, however, the fact that certain firms were competent could be traced to their TC efforts rather than to size *per se*. In the case of garment, wood-working or most metal-working technologies, where the size threshold for competence was relatively low, there were still large differences in competence between firms of similar sizes. This suggests again that *technical competence was directly traceable to deliberate investments in TC development.*[17]

2. *There is no significant difference between the age in production* of the two groups. The technological learning process is not a simple function of years of experience, but more the result of a deliberate investment in creating skills and information. The ability to undertake this investment is dependent on several factors apart from age. It is interesting to note, however, that only three of the competent firms were formed after the start of the economic recovery programme of 1983 which started the liberalisation process.

3. *Performance indicators* such as growth and capacity utilisation in capable firms are significantly better for technologically capable than other firms. This is not unexpected, but it is encouraging that the technological-capability measures, which were not based on performance measures but on direct observation of technological functions, are related to it.

4. *Ownership does not matter.* Four of the firms are foreign-controlled (two being part of large MNCs); three are owned by non-African settlers, and the remaining six by local Africans. The division of the sample firms into African and other (foreign or local non-African) fails to show statistical significance. Though the mean for non-African ownership is higher in capable firms, the T-statistic fails to reach acceptable levels. This seems surprising at first sight, since there is a general presumption that MNCs would have greater TCs than local firms in a less industrialised country like Ghana. The reason is probably the small size and purposive nature of the sample, but it may also lie in the fact that existing levels of technological capabilities reflect the legacy of decades of relative isolation and hardship. Even MNCs have to make do with the base of skills that is generally available: thus, in the longer term they may well develop better capabilities than local competitors but these capabilities may not match those of their affiliates in countries with higher levels of education, training and management experience.

5. *The average wage is significantly higher* in capable firms than in other firms (the means are $65 and $44 per month respectively). This may be due to a number of factors, such as differences in size, capital intensity, labour-market distortions or location between the firms. It may, however, also indicate that capable firms employ workers with

higher skill levels, give more training and then offer higher wages to retain workers; or are more productive for given skill levels for other reasons. The data do not allow these different hypotheses to be tested properly, but there is some evidence (see below) that the more competent firms do have higher skill levels that are related to their investments in TCs.

6. *Competent firms have much higher levels of education for the entrepreneur*[18] than other firms. The mean comes to 17.1 years of education in the former and 11.7 in the latter, and the difference is highly significant statistically. This suggests that education adds to the competence, 'vision' and organisational abilities of entrepreneurs, and is explored further below.

7. *Competent firms also have much better educated production managers*. This difference is also highly significant, with the mean being 15.5 years of education for competent firms and 8.9 years for other firms. This indicates that it is not just the 'vision' of the entrepreneur that matters, but that a technically competent production manager is also needed to catalyse the learning process (the role of the technological catalyst is discussed in the next section).

8. *Competent firms employ more technical manpower*. They have a significantly higher proportion of engineers and technicians in their workforces than other firms (6.7 per cent of total employment compared with 2.8 per cent). They also have larger proportions of employees in quality control and maintenance (6.5 per cent and 1.8 per cent) than other firms. This shows the importance of having adequate numbers of technically qualified personnel who can absorb new technologies, and of paying adequate attention to certain vital process functions. However, the employment of engineers by itself does not show any statistically significant difference.

Technological competence and human capital

The relevance of 'human capital' to technological competence and development is universally accepted in the literature. However, human capital may have many ramifications, each of which should be considered separately. A firm has a stock of skills given by the background and training of the entrepreneur or business leader, the production manager (who is generally the most important person, after the entrepreneur, in deciding the technical strategy and progress of a firm), and other technically qualified personnel hired from the labour market (locally or abroad). In addition, it has workers of different levels of quality and education. Over time, it adds to this stock by investing in training its employees, in-house or externally (locally or abroad); it also loses skills as employees leave the firm to set up on their own or join other firms. These broad components of human capital are considered separately below.

Entrepreneurs and production managers

As noted above, the level of education of entrepreneurs and production managers of the technologically capable firms is significantly higher than in other firms. Here we explore in greater detail the characteristics of the sample entrepreneurs and the production managers, starting with a brief description by industry.

In the *garments* industry, the average age of the entrepreneur is 62, and none has university education. Most are secondary-school graduates, while two have had vocational training in dress-making. It has served as the entry into manufacturing for some of the local non-Africans who were required to invest in industry under the old regime. This rather simple educational background is in keeping with the nature of the technology involved, especially for garment making. Nevertheless, a comparison with Sri Lanka suggests that garment entrepreneurs in Ghana have relatively low levels of formal education. Table 7.4, with the educational background of nine Sri Lankan entrepreneurs, shows that two have MBAs from abroad and three are chartered accountants. The low educational levels of Ghanaian entrepreneurs may have contributed to the weak supply response of this industry. It is, however, difficult to relate technological performance in this subsector to anything because of the generally declining state of all the firms.

In *food processing*, where the technology is far more demanding and the sample firms are larger, the background of the entrepreneurs/ CEOs tends to be much more impressive in educational terms. Five have university degrees, of which three are in chemistry or food technology from universities in developed countries. Two are secondary-school graduates, and have worked their way up in their firms. The best firms in the sample all have highly trained leaders.

In *wood working*, of the seven firms on which this information is available, the average age of the entrepreneur (46 years) is much lower than in textiles and garments, but the general level of education is higher, though not as high as in food processing. There are only two entrepreneurs with university degrees – these are the heads of the two firms classified as technologically capable. The others are primary-, middle- or secondary-school graduates. Of these, two started as traditional carpenters.

In *metal working*, the average age of the entrepreneur (46 years) is the same as in wood working, and the level of entrepreneurial education is somewhat higher. Of the eight entrepreneurs on whom information is available, five are university graduates, all trained abroad. Four of these have engineering degrees, one has a management degree.

It is interesting to look specifically at the characteristics of the entrepreneurs of the six capable *African-owned firms*, two in food processing, one in furniture and three in metal working. Of these, the flour mill shows the capability to manage well the transfer and absorption of technology, while the fruit-processing firm shows more innovative capabilities based on the application of scientific knowledge by the entrepreneur. The furniture firm Peewood shows good management skills and is able to buy technology from abroad. The metal-working firms Domod and Alugan show good mastery of their process technologies and the ability to find economical sources of equipment. SIS Engineering shows good capabilities in all these activities as well as some product-design ability, though not at an advanced level that would require formal R&D.

Table 7.2 shows the age, educational and work experience of these African entrepreneurs. It is clear that these entrepreneurs are *relatively young and highly educated*. Technical education *per se* is not a distinguishing feature, though in the case where it does exist it is a valuable asset to the firm. Most of the entrepreneurs are from a business-studies background, and nearly all have some experience of working in a business, of which three have experience overseas (generally in the same line of business as their present one).

These characteristics have interesting, and potentially important, implications for TC development. Entrepreneurial success among Ghanaians is clearly associated with high levels of education. This may simply reflect that better educated entrepreneurs have better access to segmented factor markets and official favours. It may, on the other hand, imply that education is associated with qualities that are conducive to technological acumen: such as analytical and organisational skills, an appreciation of technological factors, the ability to seek out necessary information and the relevant professionals, and a willingness to try new methods and technologies. There is evidence of market segmentation, but on the whole we find support for the hypothesis that education provides real benefit to technological effort (aided in some cases by relative youth).

Table 7.2   Characteristics of entrepreneurs of technologically capable African firms

Firm	Age of Entrepreneur	Highest educ. qualification	Subject of qualification	Previous work experience
Golden Spoon	n.a. (a)	Chartered acc. (L)	Accountancy	In other flour mill (L)
Astek	n.a. (b)	PhD (F)	Chemistry	Standards Board (L)
Peewood	42	BA (F)	Management	Furniture firm (F)
Domod	46	BA (L)	Busin. acc.	Alum. marketing (F)
Alugan	35	MSc (F)	Management	Employed (F, L) (c)
SIS Engineering	46	BSc (F)	Plant engineering	Univ. lecturer (d)

(a) Relatively young, probably in 40s. The firm was started by a group of similar young people.
(b) Probably in his 50s.
(c) The present entrepreneur was educated and worked in the USA before taking over father's business.
(d) The entrepreneur was a lecturer in the University of Science and Technology, and was educated in the UK, where he also worked for some time.
F = Foreign, L = Local.

Source: Lall et al. (1994).

Work experience in general has obvious benefits for the accumulation of technical know-how and institutional and marketing skills. Work experience overseas probably gives exposure to a broader range of experience and techniques. These associations are not at all surprising, but it is interesting to have them show up so clearly in this sample.

The implication is clearly not that all African entrepreneurs have to be modern, well educated, young people with work experience. There are always exceptional entrepreneurs who are 'born and not made', and rise above the constraints of low educational status to use the skills of others in building up successful businesses. However, it is still likely to be true that success in modern industry is facilitated by the cognitive, social, technical and other skills imparted by education.

The background of the *production managers* is directly relevant for explaining the ability of firms to develop technologically. In *garments*, the two most capable firms, Nitra and Overseas Knitwear, have production managers – an expatriate (German) textile engineer in the former and the owner's second son in the latter. Nearly all the technological capabilities that exist in both firms are directly traceable to these production managers (they are the technological 'catalysts' discussed later). The German has also introduced the CAD system to Nitra, which is used by a Ghanaian trained by him and has proved a competitive advantage to the firm.

In the *food-processing* industry, each production manager is university trained, one with postgraduate education in food science. Three are mechanical engineers: there are two expatriates in MNCs, while the third operates with a local biochemist. One firm has a local non-African; the others are all Ghanaian Africans. The noteworthy one is the production manager of Picadilly, who used his mechanical engineering skills to transform the obsolete machinery into a well-functioning plant.

In the *wood-working* industry the general level of education of production managers is not very high. The two smallest firms do not have a production manager at all. One has a production manager with only primary education. Another has a secondary graduate with no training in wood working. According to the available information, only the two technologically capable firms have production managers with diplomas in joinery and wood working.

In *metal working*, two firms (both in structurals) have no production managers. Of the others, only an Indian-owned steel mill has a production manager with a graduate engineering degree in metallurgical engineering, probably necessary in such a complex technology. The others all have diploma holders in mechanical engineering, except for one, which has a secondary-school graduate with previous experience in the industry (it has an engineer in charge of maintenance and QC). This is generally in conformance with the relatively simple nature of the technologies employed by these firms.

It may be relevant at this point to discuss the role of the *technological catalyst* in the relatively small and low-technology firms that populate the Ghanaian industrial sector. The 'catalyst' is an individual whose efforts and knowledge are critical to the technological upgrading of the enterprise. In most of the competent sample firms there was usually someone who played this role, generally the production manager or equivalent who worked closely with the entrepreneur or else was given a free hand to upgrade the technology of the enterprise. In small, newer, less mature enterprises, the existence of a catalyst seems to be essential to technological development.[19] In some cases it is the entrepreneur himself, setting up in business to exploit his skills or innovations. In others, it is someone selected by the entrepreneur to take the technological lead: in this sense, it is a reflection of the entrepreneur's education and vision.

Technical manpower

Table 7.3 shows the breakdown of employment by various technical qualifications, *including* the entrepreneur and the production manager. The categories used are *scientists, engineers and technicians* (together

Table 7.3   Technical manpower indicators and training in Ghanaian firms

Firm	Engrs. & techs. (a)		Engineers only (b)		QC personnel	External training (c)
	No.	% of emp.	No.	% of emp.	% of emp.	% of emp.
Textiles and garments						
Nitra	1	1.1	1	1.1	3.3	0
Adas	1	1.4	0	0	0	0
Awura Abena	1	2.4	0	0	0	0
Terrycott	1	3.3	0	0	3.3	0
Overseas Knitwear	1	4.2	1	4.2	0	0
Thadani	0	0	0	0	0	0
Wiredu	1	7.7	0	0	0	0
Mutual Union (c)	0	0	0	0	0	0
Food processing						
Nestlé	130	20.0	n.a.	n.a.	2.0	1.2
Cadbury	10	4.1	1	0.4	2.8	2.4
Gihoc Cannery	8	3.5	1	0.4	1.8	3.5
Picadilly	2	0.9	1	0.5	0.9	0
Golden Spoon	11	5.5	1	0.5	2.0	0
Fan Milk	15	8.7	3	1.7	2.9	0.6
Astek	4	5.0	1	1.3	5.0	5.0
Wood working						
Furnart	7	2.6	2	0.7	n.a.	0
TMG	2	0.9	2	0.9	0	0
Peewood	3	2.0	0	0	0.7	0
Ashanti Furniture	2	1.5	0	0	0	1.5
Barhat	3	4.6	0	0	0	0
Kyere	0	0	0	0	0	0
Progressive	0	0	0	0	0	0
Amoo	0	0	0	0	0	0
Metal working						
Tema Steel	17	4.0	3	0.7	n.a.	0
Domod	5	3.3	1	0.7	0	0
Alugan	10	14.1	0	0	1.4	4.2
UTC	1	3.3	0	0	3.3	3.3
Agbemskod	2	7.1	0	0	0	7.1
Addoh	0	0	0	0	0	0
SIS Engineering	3	15.8	1	5.3	0	0
Suame Foundry	3	18.8	1	6.3	0	0
Halaby	1	7.7	1	7.7	0	0

Notes: (*a*) All degree and diploma holders. (*b*) BSc degrees in different types of engineering.
(*c*) External training refers to numbers of employees sent on courses in Ghana and abroad in 1990–91.

Source: Lall *et al.* (1994).

and separately). The employment of scientists, engineers and technicians was found in the previous section to be significantly larger in
technologically capable than in other firms. The firm-level figures are
now considered by industry, and comparisons are drawn with other
developing countries wherever possible.

In *textiles and garments*, each of the firms has one technical person, with the exception of the dead firm and a small firm. Not surprisingly, none has a science degree, since this activity has no need for such training. However, only two firms have an employee with an engineering degree – and these are the best firms in the sample. The others have non-degree-level technicians, essentially to service the sewing machines. The total level of technical employment is low, even by the standards of this simple industry.

A comparison with Sri Lanka, a relative newcomer to the industry and not as advanced in garment quality or technology as the East Asian NIEs, can illustrate this point. Table 7.4 shows the employment of engineers and technicians in a sample of Sri Lankan firms.[20] Each of the ten garment firms (large operations with 330 to 1900 employees each) has at least 10 technical personnel, six firms have more than 20 technical personnel and two firms have more than 40 technical personnel. In more sophisticated operations, say, in Hong Kong and Taiwan, this figure is likely to be higher. The technical level of the Ghanaian garment industry, as measured by its use of engineering and technical personnel, is very low.

In *food processing*, the picture is different. Nestlé claims to have 20 per cent of its employees technically qualified, the absolute number exceeding the rest of the sample put together. The dairy-products affiliate comes next, with nearly 9 per cent of its employees with technical qualifications. Cadbury is relatively low considering the nature of its technology (which is very similar to Nestlé's). The lowest is the biscuit maker (Picadilly), owing to the simpler nature of its product.

In general, these data correspond with the relative performance of the firms within their respective technological segments. Nestlé, for instance, is distinctly a better performer in the market than Cadbury, and the two multinationals show very different propensities to invest in human capital. Picadilly does not really need any highly qualified technicians, but its access to an engineer (the production manager) allows it to perform very well.

In the *wood-working* industry, none of the firms has any scientists, and only two have any engineers. Technicians are found in four firms; the three smallest firms do not have a single technically qualified person (including the entrepreneur). In the absence of data on other countries, it is difficult to assess how Ghanaian firms fare in relative terms.

In *metal working*, there are no scientists, and relatively few engineers. Of the seven engineers in the industry, three are in Tema Steel, all recently imported from India. The others are distributed over four

Table 7.4 Entrepreneurship, technical manpower and training in Sri Lankan firms

Firm	Entrepreneur's highest level of education	Entrepreneur's educational specialisation at tertiary level	Engineers and technicians		Engineers only		QC manpower	% Trained externally
			No.	(% of emp.)	No.	(% of emp.)	(% of emp.)	(% of emp.)
		Garments						
Alliance	Secondary	None	17	5.2	0	0	4.6	1.5
Cadillac	Chartered acc.	Acc.	27	2.7	1	0.1	9.5	3.4
Colmans	Secondary	None	12	1.8	0	0	2.8	3.8
Dial	Chartered acc.	Acc.	36	2.4	11	0.7	5.9	1.0
Eskimo	MBA (abroad)	Busin. studies	56	6.2	2	0.2	4.5	2.2
Gartex	n.a.	n.a.	11	1.8	1	0.2	3.1	2.0
Hirdaramani	Secondary	None	20	2.3	0	0	3.5	0.7
Kundanmal	Chartered acc.	Acc.	25	2.5	5	0.5	8.4	1.7
Smart	MBA (abroad)	Busin. studies	43	2.3	7	0.4	3.2	1.0
Translanka	Secondary	None	15	1.7	3	0.3	3.3	0.2
Average			26.2	2.9	3.0	0.2	4.9	1.8

(continued on page 188)

Table 7.4 continued

Firm	Entrepreneur's highest level of education	Entrepreneur's educational specialisation at tertiary level	Engineers and technicians		Engineers only		QC manpower	% Trained externally
			No.	(% of emp.)	No.	(% of emp.)	(% of emp.)	(% of emp.)
		Light engineering						
Agro	Secondary	None	5	6.7	1	1.3	4.0	1.3
Browns	Secondary	None	53	4.4	40	3.3	0.3	0.6
Commercials	Chartered acc.	Acc.	37	2.6	7	0.5	0.1	0.3
Premier	Secondary	None	15	3.6	3	0.7	0.5	1.0
Walkers	Secondary	None	38	3.2	13	1.1	0.1	0
Acme	BSc	Mechan. eng.	12	6.8	5	2.8	5.6	7.9
Alpha	Secondary	None	6	3.0	2	1	2.0	0.5
Elsteel	BSc	Mechan. eng.	8	5.3	4	2.7	5.4	5.3
Metalix	BSc	Electri. eng.	7	4.4	4	2.5	3.1	0
Swedelanka	n.a.	n.a.	3	7.5	1	2.5	3.6	4.0
Average			18.4	4.7	8.0	1.8	2.5	2.1

Source: Wignaraja (forthcoming)

Table 7.5   Share of engineers in employment in other developing countries

Country	Firm	Year	Ownership	Employment	Engineers as % emp. (a)
Korea	Daewoo Heavy Industries, Diesel Engine Branch	1978	Local	1181	14.0
	Kolon	1984	Local	4132	5.6
	Daesung Electric Industries Co.	1993	Local	946	6.6
	Iljin Industries Co.	1993	Local	422	14.2
	Dong A Corporation	1993	Local	283	11.3
	Sangshin Brake Industry Co.	1993	Local	537	7.4
	Poonsung Precision Co.	1993	Local	857	3.5
Malaysia	Eng Group	1993	Local	230	4.3
	Inventec	1993	Foreign	2300	5.2
	Panasonic	1993	Foreign	996	8.0
	Motorola	1993	Foreign	5000	4.4
	Sony	1993	Foreign	4900	12.7
India	Associated Babcock	1982	Local	5000	10.0
	Hindustan Machine Tools	1981	Local	25000	5.1
	TELCO	1982	Local	39500	4.3

Note: (a) The averages of engineers to total employment in the three countries are: Korea (8.9 per cent), Malaysia (7.6 per cent) and India (6.5 per cent).

Sources: Korea from Enos (1992) and Yun (1994), Malaysia from Lall *et al.* (1994) and India from Lall (1987).

firms, including 'capable' firms like Domod and SIS Engineering. In the very small firms, the figures are difficult to interpret, because the presence of one person shows up as a large percentage. If these are ignored, the employment of *engineers* in the Ghanaian metal-working firms is *under 1 per cent of total employment*. This may be compared with some figures for the employment of engineers by large firms in engineering products in Korea, Malaysia, India and Sri Lanka (Table 7.5). The average percentages of engineers in total employment in the other countries are: Korea (8.9 per cent,) Malaysia (7.6 per cent), India (6.5 per cent) and Sri Lanka (1.8 per cent).

These figures should not be taken as direct indications of skill gaps in Ghanaian firms, since the technological level of the firms in the other countries is far higher than those of the Ghanaian sample. However, this is less true of Sri Lankan firms, which are relatively small and in simple technologies, and so more comparable to the Ghanaian sample. This comparison is particularly revealing, since it is difficult otherwise to establish whether Ghanaian firms have adequate technical manpower to achieve efficiency. Moreover, even the data on India, Korea and Malaysia are useful to illustrate the kind of skill upgrading that may be needed by Ghanaian industry if it is to enter more complex engineering activities.

Training

The level of employee skills of any firm is affected by its training strategies and by turnover of its employees. Training can take three main forms: (1) apprenticeship, which generally refers to training given to a young entrant who knows little about the skill in question, and who learns by working alongside an experienced worker; (2) on-the-job training, which generally refers to further hands-on skills imparted to a person who already has some theoretical knowledge of the work; and (3) formal employee training, where experienced employees are given formal training to refresh, increase or alter their skills.

It was possible to collect information on the numbers of employees sent by firms for external training, in Ghana and overseas. Table 7.3 above shows these numbers for 1991, as a percentage of total employment. External training is undertaken by very few firms (nine in total), and seems to be very low in relation to the skill needs of the activities undertaken. Again, a comparison with Sri Lanka shows that there was considerably less sustained external training by garment and metal-working firms in Ghana (Table 7.4 above). By Korean standards, the extent of external training in the Ghanaian metal-working firms is also extremely low: Yun (1994) reports that the average percentage of employees sent on external training, in total employment, in five Korean-owned metal-working firms was as high as 29.2 per cent in 1993. Of the training offered by Ghanaian firms, most seems to be in management rather than technical skills, and is sporadic rather than sustained. There are few institutions available in the country to provide industry-specific training in activities such as garments and textiles, or furniture making.[21] It is therefore difficult to judge the contribution of the external training that does take place in the sample.

There are no data available on training conducted internally by the sample firms. Few of the firms have separate training departments with separate budgets. However, it is possible to review qualitatively the internal training provided. These findings are reviewed by industry.

The *garment* end of the textile and garment industry in most developing countries does not require high levels of education among its shopfloor employees. It does, however, have to invest in training its workforce, especially where export markets, with very demanding standards, are served (Lall and Wignaraja, 1994). Only two firms in this group have training programmes: Nitra and Overseas Knitwear, the two technologically capable firms. In Nitra, training is by senior employees, under the close supervision of the German production manager. New recruits get 3–4 weeks of training. All staff get occasional half-day courses by the production manager on quality control, clearly an important input into the quality edge that the firm enjoys over local competitors. This sort of personal and systematic attention to training is lacking in other firms.

All the *food-processing* firms provide on-the-job training for employees. Nestlé has the best formal training programme (probably the best of all the sample firms). Apart from the programmes for new recruits, Nestlé has annual training for all employees, coordinated by an expatriate training officer. Technical training is conducted on the job by expatriate trainers. Upper management is sent to Nestlé's training centre in the parent's home country for courses in management and marketing. In general, the level of investment in human-capital development in the other food-processing firms seems to be geared to meeting basic production needs rather than to upgrading the stock for coping with technical change and competitiveness.

In *wood working*, Furnart sets the highest standards of training. It does not take apprentices, but has an extended in-house training programme for production skills that runs over 3 years, and foreign consultants will provide QC training periodically. By contrast, most other firms take recruits with minimal levels of schooling and give them apprenticeship training. The training period varies between 2 and 6 years. These firms also try to hire apprentices trained by other firms. They expressed no dissatisfaction with this system of recruitment and training, perhaps reflecting the low levels of skills that had to be imparted.

In *metal working*, Tema Steel (too complex an activity to rely on apprentices) provides considerable on-the-job training. The new owners hired new recruits from middle schools, and provided intensive on-the-job training by the expatriate personnel. The level of skills was found to be well below that in India and it is felt that more formal on-the-job training systems need to be installed after the task of refurbishing the plant is completed.

Most other firms recruit from primary schools and give apprenticeship training along traditional lines as in wood working (see above).

Interestingly, they did not want technical-school graduates for higher-level skills; their owners were suspicious of technical-school graduates, and found their training 'too theoretical' and their wages too high. This may be a reflection of their lack of formal training and their traditional craft background.

To sum up, there are ten firms in the sample (six in wood working and four in metal working) that use apprenticeships for training their workforces. Only two of these, Peewood and SIS Engineering, have been classified as technologically competent. Both have made adaptations to the traditional system to enhance its training potential. SIS Engineering recruits some workers with technical training for supervisory work along with primary school-leavers for the shopfloor; it supplements apprenticeship with follow-up on-the-job-training.

In general, the apprenticeship system seems well suited to the transmission of fairly simple manufacturing skills to workers with minimal formal education, with little change over the generations. It is less suited to training for the skills needed for modern manufacturing, where completely different types of skills from those possessed by traditional craftsmen may be required, and where a considerably higher level of education is necessary to operations. Even in activities where there is a role for traditional skills like metal and wood working, an upgrading of the apprenticeship system to encompass more formal education is called for.

POLICY IMPLICATIONS

The evidence suggests that Ghanaian industry operates with relatively low levels of human capital and invests little in upgrading that capital. There are differences between firms, and education and training show up as important determinants of technological competence. While this confirms expectations as well as the received wisdom from other countries, it is important to go beyond this. The most striking feature of the findings on Ghana is *how low overall levels of skill and competence in manufacturing industry are*. This means that technological and managerial capabilities remain well below the levels needed for Ghana to develop a dynamic industrial sector and to mount a sustained export drive based on new manufactured products (including higher value-added processing of local natural resources). This has important implications for the theme we started with: the weak 'supply response' of Ghana (and Africa as a whole) to policy reforms and import liberalisation.

Let us look at the implications of this, not from the viewpoint of education policies but from that of *industrial development and industrialisation strategy*. To some extent, Ghana's lagging competitiveness and dynamism clearly reflect the weak base of technical, managerial and workforce skills. This is not, of course, a uniquely Ghanaian problem. The whole of sub-Saharan Africa faces severe shortages of skills, especially of technical skills relevant to modern industry.[22] For instance, in 1990 Korea alone had 411,000 university students enrolled in technical subjects, compared with 111,000 for sub-Saharan Africa as a whole. And this does not take into account the quality of the technical training in Africa, which is likely to lag well behind the NIEs. Yet Korea was at roughly the same level of income as Ghana some three decades ago when the industrialisation process started.

The importance of skills to industrial competitiveness is of course now widely recognised, and there is considerable effort to build and improve education and training structures in developed countries as well as in developing ones. In sub-Saharan Africa, however, this problem is surprisingly absent from the policy discussion in the context of policy reform and SAPs. This is despite the attention paid to 'capability building' in Africa in many pronouncements by multilateral institutions.

Liberalisation does not address any of the skill shortages that may be affecting the efficiency of African industry, yet many existing industries might become competitive if their human resources were improved. It is important to note that a certain amount of capability development has already taken place in many industrial activities; this is a valuable resource that should be conserved and built upon rather than dissipated by a shock therapy that leads to massive deindustrialisation. Technological capabilities reside in *groups* of skilled and experienced persons rather than in individuals, and the destruction of enterprises means that the stock of accumulated knowledge is effectively destroyed even if the individuals concerned stay in the country. The design of SAPs should therefore include education and training as an *integral* part of the reform and restructuring process.

The provision of *specialised training* to industry is an important area of supply-side support in the process of structural adjustment.[23] These services are weak in much of Africa, and enterprises themselves (apart from the major multinationals) invest little in training their employees in modern technologies. In Africa there is a great shortage of experienced trainers to staff and manage industrial training systems, and this is the first bottleneck that governments should address in the context of adjustment. SAPs should pay explicit attention to the need for foreign trainers to teach local trainers in the most pressing skill needs of industry, and to the need to set up viable local training systems in the longer term. The problem is once more the shortage of human resources, and has to be tackled at source.

The levels of skills and capabilities needed by industry also need to be enhanced by providing *technical extension services* to industry, especially to small and medium enterprises (SMEs). Those technical extension services that exist in much of Africa are largely ineffective in providing the kind of inputs that enterprises need in order to upgrade to competitive levels. The Ghana study showed clearly how the numerous institutions that had been set up to help industrial technology contributed practically nothing to enterprises (see Lall *et al.* 1994). At the same time, the experience of East Asia shows that the governments there played a much larger role in upgrading the technologies used by their enterprises than interventionist governments in Africa, and that many of these services were extremely demanding in resources.[24] It is also clear that such services should have formed intrinsic parts of SAPs if they were to have any long-term effect on industrial upgrading.

So much for supply-side support to improve industrial skills and capabilities. Let us conclude with the implications for reform to the *incentive system*. The analysis of Ghanaian capability development suggests that a gradual and controlled process of opening up, accompanied by a *strategy* of industrial restructuring and upgrading, is to be preferred to a rapid and sweeping exposure to market forces envisaged by the 'ideal' SAP. The speed of liberalisation should be based on a realistic assessment of which activities are viable in the medium term, with the process geared to the learning and relearning needs of various activities. In this entire process the government should retain powers to guide resource allocation, but in a clear and transparent manner, and with strict requirements of capability development. It involves close monitoring of the progress of liberalisation, and it requires that the government is able to address the supply-side needs of industries along with allowing a phased process of liberalisation.

To recommend a more gradual and nuanced strategy of liberalisation is *not* to suggest that Ghanaians should simply slow down the adjustment process. They should instead *actively prepare* for it in the grace period provided. There is, however, a strong risk of *government failure*. The launching pad of any reform must be improvements in government capabilities themselves. Evidence suggests that these can be improved by training, reorganisation of the civil service, better performance incentives and monitoring, and greater insulation from the political process.[25] Without such capability development, *even the market-friendly strategies recommended by the World Bank have little chance of success.* At the same time, it is not recommended that Ghana attempt the detailed and pervasive interventions practised in a country like Korea; this does impose tremendous demands on the government and runs very high risks of hijacking and abuse. The correct form and level of selectivity that can be managed by particular governments at particular times is itself a subject that deserves close study, but is ignored by donors who insist that all selectivity is undesirable.

REFERENCES

African Development Bank (1994), *African Development Report* (Abidjan: African Development Bank).
Bell, M. and Pavitt, K. (1993), 'Technological Accumulation and Industrial Growth: Contrasts between developed and developing countries', *Industrial and Corporate Change*, 2(2): 157–210.
Enos, J. (1992), *The Creation of Technological Capabilities in Developing Countries* (London: Pinter).
Fagerberg, J. (1988), 'International Competitiveness', *Economic Journal*, 2: 355–74.
Fontaine, J. M. (1992), 'Bias Overkill: Removal of anti-export bias and manufacturing investment: Ghana 1983–89', in R. Adhikari, C. Kirkpatrick and J. Weiss (eds), *Industrial and Trade Policy Reform in Developing Countries* (Manchester: Manchester University Press).
Hussain, I. and Faruqee, R. (eds) (1994), *Adjustment in Africa: Lessons from the Country Studies* (Washington, DC: World Bank).
Katz, J. M. (ed.) (1987), *Technology Generation in Latin American Manufacturing Industries* (London: Macmillan).
Lall, S. (1987), *Learning to Industrialize* (London: Macmillan).
Lall, S. (1992.a), 'Structural Problems of African Industry', in F. Stewart, S. Lall and S. Wangwe (eds), *Alternative Development Strategies in Sub-Saharan Africa* (London: Macmillan).
Lall, S. (1992.b), 'Technological Capabilities and Industrialization', *World Development*, 20(2): 165–86.
Lall, S. (1993.a), 'Understanding Technology Development', *Development and Change*, 24(4): 719–53.
Lall, S. (1993.b), 'Trade Policies for Development: A policy prescription for Africa', *Development Policy Review*, 11(1): 47–65.
Lall, S. (1994), 'Industrial Policy: The role of government in promoting industrial and technological development', *UNCTAD Review*: 65–89.
Lall, S., Navaretti, G. B., Teitel, S. and Wignaraja, G. (1994), *Technology*

and Enterprise Development: Ghana under Structural Adjustment (London: Macmillan).

Lall, S. and Wignaraja, G. (1994), 'Foreign Involvement by European Firms and Garment Exports by Developing Countries', *Asia-Pacific Development Journal*, 1(2): 21–48.

Lazonick, W. (1993), 'Learning and the Dynamics of International Competitive Advantage', in R. Thomson (ed.), *Learning and Technological Change* (London: Macmillan).

Leechor, C. (1994), 'Ghana: Forerunner in adjustment in I. Hussain and R. Faruqee (eds), *Adjustment in Africa: Lessons from the Country Studies* (Washington, DC: World Bank), pp. 153–93.

Lucas, R. E. B. (1994), 'Impact of Structural Adjustment on Training Needs', *International Labour Review*, 133(5–6): 677–94.

Mosely, P., Harrigan, J. and Toye, J. (1991), *Aid and Power: The World Bank and Policy-Based Lending* (London: Routledge).

Steel, W. F. and Webster, L. (1992), 'How Small Enterprises in Ghana have Responded to Adjustment', *World Bank Economic Review*, 6(3): 423–38.

Stein, H. (ed.) (1995), *Asian Industrialization and Africa: Studies in Policy Alternatives to Structural Adjustment* (London: Macmillan).

Wignaraja, G. (forthcoming), *Trade Policy, Technology and Export Performance: Sri Lanka's Liberalization Experience* (London: Macmillan).

World Bank (1991), *World Development Report 1991* (New York: Oxford University Press).

World Bank (1993), *The East Asian Miracle: Economic Growth and Public Policy* (New York: Oxford University Press).

World Bank (1994), *Adjustment in Africa: Reforms, Results, and the Road Ahead* (New York: Oxford University Press).

Yun, M. (1994), 'Effects of Sub-Contracting Relationship on Technological Capability of Korean Automotive Suppliers' (Doctoral thesis, in progress, Oxford University).

8 Foreign Direct Investment Policies in the Asian NIEs

INTRODUCTION

Two broad inter-related policy issues arise for developing countries in the context of international investment.[1] The first is whether and how much FDI to allow in, i.e. if one should exercise selectivity in letting in MNCs. The second is, having allowed in FDI, whether to intervene selectively in the operations of MNCs, setting conditions for their operations, and targeting investments of higher 'quality'. Both forms of intervention may be desirable if there is a perceived divergence between private and social returns from MNC activity in free markets. The first set of issues is determined by the costs and benefits of FDI to the developing host country as compared with alternative ways of accessing capital, technology and skills. The second is determined by market failures in domestic (and foreign) markets which guide MNC activities and which may be altered to obtain larger social benefits for the host economy.

(i) Take the *entry and extent of FDI*. The literature on international investment attributes the existence of MNCs to the presence of failures in the markets for the intangible assets that constitute their 'ownership advantages' (Dunning, 1988). Without such advantages there would be no reason for MNCs to come into being: the essence of transnationalisation is the internalisation of imperfect intermediate markets. This by itself has no particular policy implications. As long as the internalised markets of MNCs are efficient, not just for the firm, but also for host economies, there is by definition no divergence between private and social interests. If, however, there is such divergence, the imposition of (privately efficient) internalised markets by MNCs may not always be beneficial for the host economy.

MNCs are among the most powerful means available for transferring modern technologies to developing countries and overcoming obstacles to their utilisation. By virtue of their large internal markets for capital, skills, technology and information, they face fewer market failures than local firms. In most circumstances, therefore, there is no reason to

197

place restrictions on their entry – their presence can only benefit local productivity and competitiveness. Moreover, since MNCs are at the forefront of innovation, their presence provides an effective means of keeping up with technical progress. Their established brand names, global marketing presence and international flows of information all add to their technological advantages.

What case can there be for exercising selectivity on FDI? Three reasons can be found in the development literature.

First, there is an important distinction between the transfer and utilisation of production technologies and the transfer and development of more complex design, development and innovative capabilities. Innovative activity by MNCs tends to be concentrated in a few developed countries, because of the location of management and decision-making centres, the availability of advanced and specialised technical skills, large local markets, linkages with established suppliers and buyers, closeness to advanced science and technology institutions, and proximity to central decision making. The upgrading of capabilities in developing countries to the levels needed for high-level technological activity generally involves high learning and other costs which foreign investors tend to be unwilling to take. In less developed economies MNCs may hold back the development of innovative capabilities while enhancing production capabilities; it is mainly the more advanced industrial countries that can attract and fully benefit from the transfer of innovative capabilities by MNCs (Dunning, 1991). Thus a passive reliance on MNCs to upgrade and deepen technological capabilities may take a very long time to bear results.

Secondly, the development of high-level capabilities in local firms may be more beneficial than a similar development within MNC affiliates. This would be the case where technological development by local firms leads to greater spillover benefits and linkages (to local suppliers and institutions) within the host economy.

Finally, a strong MNC presence in industry, while stimulating local competitors to be more efficient in their production, can inhibit them from deepening their technological capabilities. Because of the higher risks and longer learning periods involved in creating a design and development capability, local firms exposed to full MNC competition may prefer to import foreign technologies proven and 'ready made' from overseas rather than invest in their own R&D capabilities.

There may, therefore, be deficiencies in technological deepening in relying *passively* on the transfer of technology through MNCs, leading to a relatively static pattern of specialisation as far as capability development is concerned. But it may be in the interest of the industrialising country to promote technological deepening. Technological deepening would allow countries to import and absorb new technologies more economically, enter into more advanced activities, keep abreast of new developments, develop new products and processes, and better utilise local resources and linkages. The argument for restricting reliance on internalised means of technology transfer to induce technological development is rather similar to the case for intervening to promote comparative advantage by fostering infant industries, and rests on the remedying of similar market failures in information, capital, technology and other markets (of course, this is a clear case where the static market-failure approach seems most inadequate in dealing with policy issues, since interventions can take an economy into a completely new realm of competitive advantage in a dynamic sense).

The deepening of local technological capabilities is not an argument for the wholesale exclusion of FDI. On the contrary, it suggests a need for selectivity only in activities and at times when local technological development is feasible and desirable. In circumstances where the host economy is not capable of economical technological deepening, or where the technology is so closely held or advanced that local development is not possible, a reliance on FDI would be fully endorsed. Moreover, in some cases technological deepening would be achieved not by keeping out MNCs but by inviting them and influencing their activities (see below). Technological deepening can itself become a major factor in attracting higher 'quality' and more FDI: if local innovative capabilities advance, it becomes in the interests of MNCs to transfer more complex activities and R&D itself to those countries.

If it is accepted that *some* interventions are needed to speed up technological development, then we can distinguish two broad strategies for intervention to promote technological deepening:

- One would be to increase dependence on FDI, but to induce MNCs, by a mixture of incentives, rules and negotiations as well as investments in local skills and institutions, to enter activities with more complex technologies, upgrade local technological capabilities within given activities, establish closer linkages with local technology institutions and set up local R&D units.

- The other would be to adopt a more independent strategy, restricting technology imports via FDI and promoting it in 'externalised' forms (such as licensing, joint ventures or other means, in which local firms retain control and invest in deepening and extending their technological capabilities) in circumstances where this is warranted. It must be noted, however, that local ownership or control *per se* would not ensure that deeper innovative capabilities would develop (Najmabadi and Lall, 1995). Local firms might choose to remain passively dependent on imports of foreign technology and skills, and, if they had lower technical skills and managerial capabilities, and were more risk averse, might develop less technological capabilities than foreign affiliates. The development of deeper capabilities would require other *complementary* interventions to ensure that incentives existed for local firms to invest in such risky activity, the necessary skills and information were available, and capital markets were able and willing to finance the process, or local firms were promoted to a size that enabled them to internalise capital and other relevant markets.

The choice between the FDI and nationally-led strategies of technological development will depend partly on the country's political economy (some countries, for instance, are committed to open FDI policies or lack the tools of intervention or the local entrepreneurship to mount effective national technological strategies), and partly on the size and spread of the industrial sector (smaller economies with more specialised industries may prefer the FDI-led route while larger ones with diverse sectors may prefer the national route). The East Asian experience shows successful examples of both strategies.

(ii) Consider now the *second set of policy needs*, to influence or intervene in MNC activities in order to promote the upgrading of competitive and technological advantages. Market deficiencies may make it necessary, for instance, to undertake measures to strengthen a country's attractiveness to MNCs, guide MNC entry into activities conducive to industrial upgrading, and develop factor markets in ways that lead to upgrading in the 'quality' of FDI. Given the basic preconditions for attracting foreign investors, for example, governments may have to promote their countries in international investment markets and target their promotion to particular home countries or MNCs (Wells and Wint, 1990) to overcome deficiencies in information markets.

In responding to free market forces, foreign investors would focus on activities that exploited the host country's given competitive advantages rather than those that could be developed with some additional effort. The upgrading of investment activities would thus wait for the accumulation of production factors and the reflection of this in relative factor prices; even so, host countries with high wages and rising stocks of capital might not attract complex and high-skill indus-trial investments. Intervention could be used to promote the upgrading of MNC activities from simple, labour-intensive and low-technology to more complex and demanding ones, by guiding foreign entry or providing strong incentives to all investors. Intervention could also be needed to induce MNCs to deepen their technological activities in host countries, from those needed for final assembly and processing to those needed for adaptation, design, development and finally innovation. Such intervention might involve inducing MNCs to strengthen local tech-nology activities, establish closer links with local technology institu-tions and set up in-house R&D activities, or strengthening the base of local supplier firms, technical skills and technology institutions, or a mixture of both.

These are not merely hypothetical policy issues – as shown below, they guided policy makers in East Asia. The need for selective and other interventions to promote industrialisation was widely (though not universally) recognised. However, each government perceived differ-ent policy needs in line with its own strategy, identifying different market failures and adopting different solutions. Some chose to intervene very little in either the entry of MNCs or their subsequent activities; some to rely heavily on FDI but to intervene in MNC operations; and some to reduce reliance on FDI and to intervene extensively to pro-mote local enterprises and indigenous technological capabilities. All this took place in a common setting of strong export-orientation, private-sector primacy, well managed macro-economies and strong, capable governments. These factors, while providing the necessary conditions for industrial success, are not sufficient to explain the nature and ef-fects of the particular industrial and internationalisation strategies fol-lowed by the different NIEs.[2]

EAST ASIAN FDI STRATEGIES

As far as the NIEs and 'new' NIEs are concerned, there have been large variations in their propensities to attract FDI. The largest host countries for FDI were Singapore and Malaysia, relatively small econ-omies by regional standards, while the large economies of Korea and Taiwan were fairly small recipients, with the amount of FDI inflows declining in the past two to three years. Hong Kong was a relatively large recipient but with stagnating inflows; Indonesia was on a rising trend, while Thailand seemed to be stagnant or declining.

Table 8.1　Inward FDI as a percentage of gross domestic investment

Country	1981–85	1986–90	1991	1992
Hong Kong	6.5	13.6	2.3	–
Singapore	181	33.9	26.2	30.6
Korea	0.5	1.3	1.0	0.5
Taiwan	1.5	3.5	3.0	–
Indonesia	1.0	2.0	3.7	4.1
Thailand	3.2	5.9	5.6	–
Malaysia	10.8	10.6	23.9	–
Japan	0.1	–	0.1	0.2

Source: UNCTAD (1994), Annex Table 5.

These large disparities are illustrated in Table 8.1, which shows the share of FDI in gross domestic capital formation in the above countries and in Japan. It shows, in very broad terms, that the countries that developed the most diverse, deep, complex and technologically dynamic industrial sectors (Korea, Taiwan and Japan) had the least reliance on FDI. It was clearly not the lack of incomes, growth or competitive potential that led to this low reliance: the reason lay in their deliberate policies to restrict FDI inflows. Certainly, their industrial strategies were directed, among other things, at the promotion of local enterprises and the development of indigenous technological capabilities, and selectivity on FDI was one important aspect of their strategies.

This suggests that the governments of the industrially more advanced countries were seeking to exploit causal relationships between the restricted entry of FDI, the growth of domestic enterprises and the development of local innovative capabilities. However, most of the other NIEs had more modest technological ambitions and less desire to promote local enterprises. At the cost of some simplification, the group may be divided into four categories as far as FDI strategies are concerned:

- Those that followed passive open-door policies on MNCs and did not intervene in other ways to selectively promote industrial development (e.g. Hong Kong).

- Those that pursued active industrial policies and promoted local enterprises in certain activities, but adopted effectively open-door, non-interventionist policies in most export-oriented sectors (e.g. Thailand, Malaysia).

- Those that actively sought heavy MNC participation in manufacturing and did not seek to promote local industrialists, but intervened pervasively and selectively to guide and induce investors to upgrade their activities and increase local technological activity (e.g. Singapore).

- Those that selectively restricted FDI and sought to maximise reliance on externalised forms of technology transfer in the context of a comprehensive set of industrial policies to deepen the manufacturing sector, promote local linkages and increase local innovative capabilities (e.g. Korea and Taiwan, and earlier Japan). These industrial policies encompassed interventions in trade, finance, skills, technology and institution building, with strongly selective aspects to practically all interventions.

Let us describe briefly the important features of the industrial and FDI strategies of the NIEs.

Hong Kong

First, the most liberal regime, Hong Kong, was able to combine free trade and substantial inward FDI with a dynamic indigenous industrial class that was very successful in export markets. Hong Kong was, however, a very special case by virtue of its location, long *entrepôt* tradition and established infrastructure of trade and finance, the presence of large British companies (the 'Hongs'), and the influx of entrepreneurs and trained textile and metalworking engineers and technicians (with considerable learning embodied in their skills) from mainland China. This unique background allowed it to launch into export-oriented light manufacturing under free trade, but it *started and stayed with* light labour-intensive manufacturing industry, where the learning costs were relatively low and predictable. Hong Kong's success was based on the development of operational and marketing capabilities, but there was little industrial deepening and R&D growth. There was some 'natural' progression up the ladder of industrial complexity as product quality was upgraded and new products were added within existing areas of strength, but it was relatively limited in relation to other NIEs.

As wages and land costs rose, the colony relocated its manufacturing to other countries, mainly mainland China, and suffered a significant loss of industrial activity at home (during 1986–92 it lost about 35 per cent of its manufacturing employment, and the process is continuing[3]). The growth of its own manufactured exports (as opposed to re-exports) slowed down considerably, and may even have gone into decline in 1993–4; its manufacturing production is also practically stagnant. Hong Kong did not seek to 'use' MNCs in any deliberate sense, and increasingly its FDI structure is specialised in service activities geared to China. Its impressive overseas-investment performance, especially in China, is a reflection of its advanced entrepreneurial and limited technological capabilities rather than of broad industrial strengths. At the same time, the lack of a strong technology base worries the government, and it is launching initiatives such as the Hong Kong Industrial Technology Centre to selectively promote local high-technology companies.[4]

The economy is continuing to grow and prosper, but the lessons of the Hong Kong 'miracle' for the rest of the developing world are ambiguous. In view of the exceptional initial circumstances of the colony, *laissez faire* would not by itself be sufficient to lead to the Hong Kong kind of industrial or export development in typical developing countries. Furthermore, the lack of industrial deepening and the massive deindustrialisation over time follow directly from its absence of industrial policy, and in the absence of a gigantic and thriving hinterland to service a similar policy it would be deemed undesirable in other developing economies. In brief, the case does not conclusively establish a general case for fully liberal policies on trade or FDI.

Singapore

In contrast to Hong Kong, Singapore illustrates clearly the consequences of a more interventionist policy on FDI and industrial targeting combined with free trade. Singapore has half the population of Hong Kong, but has developed a far deeper industrial structure (in terms of the sophistication of production and exports) and has continued to sustain high rates of industrial and manufactured export growth despite having higher industrial wages. It has the highest reliance of any country on MNCs, and has done extremely well from it; but, unlike Hong Kong, the government targeted industries and services for promotion and aggressively sought and used MNCs as the tool to achieve its objectives.

The economy started with a base of capabilities in *entrepôt* trading, ship servicing and petroleum refining. After a brief period of import substitution, it moved into export-oriented industrialisation, based overwhelmingly on investment by multinational enterprises. Unlike Hong Kong, there was a weak tradition of local entrepreneurship, with little influx of technical and entrepreneurial know-how from China. There was a decade or so of light industrial activity (garment and semiductor assembly), after which the Singaporean government acted firmly to upgrade the industrial structure. It intervened in foreign investments to guide MNCs to higher value-added activities, and in education to create the specific high-level technical skills that would be needed.[5] The government also set up a number of public enterprises to enter activities that were considered in the country's future interest and where foreign investment was considered unfeasible or undesirable; the public sector in Singapore accounts for a substantial proportion of GDP.

Specific areas of manufacturing and services were selected for promotion, with the policy instruments including incentives that guided the allocation of foreign and local resources and lowered the cost of entry into difficult activities (by providing the requisite skills and infrastructure). Manufacturing activity was upgraded into specialised processes and products, though the base of local capabilities was important in guiding this process.[6] Such specialisation, along with the heavy reliance on foreign investments for technology and skill transfer, greatly reduced the need for indigenous technological investments (compared with Korea). At the same time, the Singaporean government mounted strong efforts to induce MNCs to establish R&D facilities there to counter the fact that the technological depth of the affiliates was still comparatively low.[7] It has had some success in this.

Korea and Taiwan

The larger NIEs Korea and Taiwan treated FDI in very different ways from the above, and also from each other. They have shown a clear preference for promoting indigenous enterprises and for deepening local technological capabilities. As such, they always assigned FDI a secondary role to that of technology import in other forms. Their export drive was led by local firms, and a series of interventions (mostly selective and integrated across product and factor markets) allowed local firms to develop impressive technological capabilities. The domestic market was not exposed to free trade; a range of quantitative and tariff measures were used over time to give infant industries 'space' to develop their capabilities. The deleterious effects of protection were offset by strong incentives (in the case of Korea, almost irresistible pressures) to export and face full international competition. Given the significance of their experience in the present context of the role of government in FDI, it is worth considering their approaches at some length.

Korea

Korea went much further in developing advanced innovative capabilities and heavy industry than Taiwan.[8] To achieve this compressed entry into heavy industry, its interventions had to be more detailed and pervasive. Korea relied primarily on capital-goods imports, technology licensing, and other technology-transfer agreements to acquire technology.[9] It used reverse engineering, adaptation and own-product development to build upon these forms of arm's-length technology imports to develop its own capabilities. Korea is one of the few developing countries that has been able to use imported technology successfully to feed into its domestic technology and to develop an independent innovative base. Its R&D expenditures are now around 2 per cent of GDP, and over 80 per cent of this comes from private enterprises, by far the highest in the developing world (and ahead of all but a handful of leading OECD countries). While not a leading innovator in the normal sense, its enterprises have considerable technological muscle, able to utilise leading-edge technologies in a variety of industries.

One of the pillars of Korean technological strategy, and one that marks it off from the other NIEs (but parallels earlier Japanese experience), was the deliberate creation of large private conglomerates, the *chaebol*. The *chaebol* were hand-picked from successful exporters and were given a range of subsidies and privileges, including the restriction of MNC entry, in return for pursuing the government's industrial strategy of setting up capital- and technology-intensive activities geared to export markets. The rationale for fostering size was obvious: in view of deficient markets for capital, skills, technology and even infrastructure, large and diversified firms could internalise many of their functions and undertake the cost, risk and long-term perspective needed to absorb very complex technologies (without a heavy reliance on FDI), further develop it by their own R&D,[10] set up world-scale facilities and market their products abroad by creating their own brand image and distribution networks. This was a costly and high-risk strategy, since the dangers of fostering giant firms in a relatively small economy are obvious. They were contained by the strict discipline imposed by the government in terms of export performance, vigorous competition between the *chaebol* (except when they were bidding for international contracts) and deliberate interventions to ensure rationalisation of the industrial structure.

The technology strategy of Korea is discussed elsewhere in this book, and it need not be discussed further here – it must be noted, however, that it is of central significance to the analysis of FDI policy. The most important point to note is that while FDI was an important input into their industrialisation, MNCs were used by the government mainly to further the acquisition of technology by local firms – a very different approach from that of Singapore. The internalised markets of MNCs were not allowed to weaken the deficient factor markets of the host economy, but were tapped in such a way that local innovative capabilities were strengthened. As these capabilities grew, FDI was allowed to play a larger role, but it never became the engine of technological or industrial development. The Korean government also undertook various measures to encourage the diffusion of technology. The government put pressures on the *chaebol* to establish vendor networks; the measures to promote subcontracting and technology diffusion to SMEs are described in the chapter above on technology policies.

Apart from the array of direct interventions to support local enterprise to develop its technological capabilities without relying on MNCs, the government provided selective and functional support by creating general and technical skills. Korea today has the highest rate of university enrolment in the developing world, and produces nearly as many engineers each year as the whole of India. While much of higher education is privately financed, the government has been instrumental in setting up universities, guiding the curriculum in the directions needed by industrial policy (and involving private business in governing universities) and regulating the quality of the education.

Taiwan

Taiwan switched to an export-oriented strategy in the 1960s, but within this it implemented a comprehensive set of industrial policies, encompassing import protection, directed credit, selectivity towards foreign investors, support for indigenous skill and technology development and strong export promotion.[11] While this resembles Korean strategy in many ways, there are important differences. Taiwan did not promote giant private conglomerates, nor did it attempt the intense drive into heavy industry that Korea did. Taiwanese industry is largely composed of SMEs, and, given the disadvantages to technological activity inherent in small size, these were supported by a variety of inducements and institutional measures in upgrading their technologies (Taiwan has perhaps the developing world's most advanced system of technology support for small and medium enterprises).

As with Korea, Taiwan used a variety of means to acquire foreign technology in support of domestic development, though with less nationalistic fervour. In the early years of industrialisation, the Taiwanese government sought to attract FDI into activities in which domestic industry was weak, and used a variety of means (below) to ensure that MNCs transferred their technology to local suppliers. 'Taiwan restricted the entry and activities of multinational companies in many ways, tightening controls as goals of technological upgrading and foreign equity investments were reached.'[12] As with Korea, FDI was directed at areas where local firms lacked technological capabilities. Where necessary, the government itself entered into joint ventures, for instance to get into technologically very difficult areas such as semiconductors and aerospace.[13] The government also played an active role in helping SMEs to locate, purchase, diffuse and adapt new foreign technologies.

MNCs were made to play an important role in the process of promoting backward linkages. In the early years, the government applied minimum-content requirements in industries like motor vehicles and consumer electronics. Over time it moved to more indirect measures to promote linkages, by giving incentives for principal firms to use local subcontractors and by improving the technological and business capabilities of SMEs.

Local R&D was encouraged by tax incentives, and skill levels were improved through sustained investments in education and training. The purchase of local equipment and entry into 'linkage-intensive' activities were encouraged by tax incentives. Taiwan has set up a science town in Hsinchu, with 13,000 researchers in two universities, six national laboratories (including ITRI) and a huge technology institute, as well as some 150 companies specialising in electronics. The science town makes special efforts to attract start-ups and provides them with prefabricated factory space, five-year tax holidays and generous grants. Since 1980 the government has invested $500 million in Hsinchu. In 1993, the Taiwanese government also announced a three-year stimulus package which included NT$40 billion (US$1.5 billion) in loans to SMEs and NT$20 billion for high-technology enterprises.

This sketch of the policies of the NIEs leads to the following conclusions:

1. Selective as well as functional interventions played a vital role in the pattern of industrial and technological development in the NIEs.

2. Governments showed an ability to devise and implement interventions effectively, partly because export-orientation imposed a strict discipline on both industry and governments and partly because of the high levels of training, adequate remuneration and political insulation of bureaucrats.

3. The nature and impact of interventions differed according to differing government objectives and political economies; however, the extent of industrial and technological deepening achieved was strongly related to selective interventions to promote such deepening.

4. FDI was treated very differently by each of the four countries and so played very different roles in their technological development. Those that wanted to promote *indigenous* technological deepening had to intervene to restrict foreign entry and to guide their activities and maximise the spillovers. Those that chose to rely on MNCs and upgrade within their global production structure had to intervene to target investors, guide their allocation and induce them to set up more complex functions than they would otherwise have done.

5. The different approaches to FDI shown by Korea and Taiwan as compared with Singapore partly reflect their objective situations in addition to their political beliefs. The options and compulsions applicable to the larger economies, with greater scope for internal specialisation and local content as well as better established indigenous enterprises, were different from those open to a small island state with weak indigenous entrepreneurship and a tiny internal market. Given the need to spread technological development more widely, the former had to take more direct steps to assist local firms.[14]

CONCLUSIONS

It has been argued here that *laissez faire* policies towards FDI were not the norm in East Asia, and that there were sound economic reasons for the kinds of interventions in investment flows seen among the leading NIEs. These interventions may or may not have involved restricting MNC entry (the normal sense in which FDI interventions are regarded); on the contrary, in some cases they entailed aggressively seeking out and attracting foreign investors. They always required functional interventions to strengthen basic factor markets and institutions, in order to upgrade competitiveness and the 'quality' of FDI inflows. This was the kind of intervention practised by Hong Kong. In other economies they entailed extensive selective interventions, aimed at upgrading technologies and technological capabilities. Two broad strategies of selective intervention were identified: the 'target and guide' strategy of Singapore and the 'restrict and exploit' strategy of Korea and Taiwan. The latter strategy had sub-elements, with Korea mounting more detailed interventions than Taiwan, with stronger ambitions to enter heavy and high-technology industry and to set up its own giant firms with ownership advantages to rival those of traditional MNCs from the developed world.

Can the relative successes or failures of these governments be explained by their ability to devise and implement a *systemic or holistic approach* to overcoming market deficiencies and creating new dynamic parameters within which markets operate? Certainly each government had a very different 'vision' of the direction its economy should take and a very different political economy in which it designed policies for the private sector. Hong Kong, at the one extreme, had no vision at all, and interfered as little as possible with enterprises, except to offer smaller enterprises subsidised technical support and export-marketing assistance and information. Korea, at the other, was driven by a vision of an advanced, diversified and nationally-owned industrial sector with an autonomous ability to undertake innovation and create its own MNCs. Singapore had a vision of an advanced economy fully integrated into the global production system, exploiting its strategic location and highly skilled population to retain a leading position in production and services. Taiwan vacillated between being like Korea and a less interventionist state, but clearly with the ambition to deepen and diversify the industrial and innovation base.[15]

It is difficult to assess to what extent these visions were consciously held by policy makers and whether their interventions in realising them were carefully thought through for coherence and applicability. There was a lot of policy learning-by-doing, and the vision probably developed over time as some policies worked and the government gained confidence. They had certain common elements in terms of promoting the competitiveness of their firms: stable macroeconomic policies, well-functioning financial systems and infrastructure, a good base of skills and the encouragement of entrepreneurial activity in a hard-working community. They developed efficient technology support systems to help small and medium-sized enterprises and to mount export drives. But it is the differences that are clearly more striking than the similarities.

The main similarities among the Tigers in their attempts to become competitive (apart from a common regional effect that may have helped their growth and policy development) are probably in meeting market failures 'in the small', i.e. in remedying the kinds of static market deficiencies where market-failure analysis is most appropriate. This applies, for instance, to investing in physical infrastructure, a base of general skills, providing technical support and export-market information. Their main differences are probably in meeting market failures 'in the large', that is in formulating the national strategies that, given their initial starting positions, changed the entire basis of their competitiveness. This is where the static optimisation approach is least satisfactory, since it does not explain why Korea chose to go into D-Ram chips, steel and automobiles, Taiwan into PCs and Singapore into specific types of producer electronics or biotechnology. These were choices made out of a large range of possibilities and were not addressing any static optimisation failure. Once the choice was made within a strategic vision, the instruments used varied in their degree of selectivity and impact.

Both the similarities and the differences were essential ingredients of the success of the NIEs. It is difficult in this chapter to assess whether there was something *unique* about their social, political, cultural and institutional settings that provided both the similarities and differences. I would like to believe that while some unique, irreproducible elements may well have existed, there are enough economic lessons that can be applied by other countries, making allowances for their administrative and political circumstances. The differences between the Tigers themselves may often be greater than between one of them and another developing country. Their experience thus does offer important and relevant lessons to other regions, developed, developing or in transition.

This chapter has not been able to explore the outward investment by these countries, though all the NIEs are active and aggressive overseas investors. Their different patterns of outward FDI, and their relationship with their industrial strategies and the development of different ownership advantages, are explored in Lall (1991). However, it is clear that their industrial policies shaped the ownership advantages of their own enterprises and allowed them to globalise in differing ways. There were common elements that support Dunning's concept of the *investment development path*, that with economic development there is a clear and predictable evolution in a country's location, ownership and internalisation advantages: from a low base of local capabilities where there is little inward or outward FDI, through the growth of inward FDI followed by outward FDI and finally a balance. The Tigers each show such a path as their own capabilities have grown over time.

However, the impact of *different* strategies has been that their own MNCs have taken very different shapes and sizes within this broadly similar evolution. Hong Kong has spawned a large number of essentially 'low-tech' investors that are relocating labour-intensive operations in cheap wage areas, though in services and real estate many are also active in developed areas. At the other end, the Korean *chaebol* have become large MNCs, very similar in competence, technological abilities and market reach to the established MNCs of the advanced countries, and globalising in a way inconceivable for the other Tigers. Thus the investment development path is overlaid by individual strategic differences.

To conclude, the East Asian NIEs provide a fascinating panorama of experience in industrial development, government intervention and treatment of FDI. What is undeniable is that their governments played a critical catalytic role in forming their competitive (or ownership) advantages in trade and industry, which then determined their participation in the global economy. The approach to FDI and globalisation was an integral part of a larger industrial strategy, and MNCs were increasingly seen as a resource which could be exploited in the national interest (an important shift from earlier perceptions of MNCs).

There is nevertheless still considerable debate about the effects of the selective industrial policies in East Asia. Furthermore, there remain doubts about the extent to which the ability to mount such interventions is present elsewhere. The conditions under which governments can exercise efficient intervention are certainly not found in many developing countries. The risk of *government failure* is so great in some cases that it may be better to suffer the consequences of market failure than to indulge in selectivity. In such cases the government should confine itself to 'market-friendly' interventions and entrust the custodian role to free markets in trade and investment. However, government capabilities are not static or given in perpetuity; they can be improved, and there are various levels of selectivity in intervention. Is it possible to gear the level of selectivity to the capabilities of governments? Is it possible to raise these capabilities by specific actions and institutional mechanisms?

As long as the development process is confronted with widespread market failures, there are good reasons that careful selective and functional interventions can speed up the development. The recent swing of opinion in favour of free markets needs to be tempered with a proper consideration of the role of government. This applies to FDI as well as to other areas of industrial policy.

REFERENCES

Amsden, A. (1989), *Asia's Next Giant: South Korea and Late Industrialization* (New York: Oxford University Press).

Amsden, A. (1994), 'Why Isn't the Whole World Experimenting with the East Asian Model to Develop? Review of *The East Asian Miracle*', *World Development*, 22(4): 627–34.

Brautigam, D. (1995), 'The State as Agent: Industrial development in Taiwan, 1952–1972', in H. Stein (ed.), *Asian Industrialization and Africa* (London: Macmillan), pp. 145–82.

Chang, Ha-Joon (1994), *The Political Economy of Industrial Policy* (London: Macmillan).

Dunning, J. H. (1988), *Explaining International Production* (London: Unwin Hyman).

Dunning, J. H. (1991), 'Multinational Enterprises and the Globalization of Innovatory Capacity' (Reading: University of Reading), Draft, p. 65.

Dunning, J. H. (1996), *Governments, Globalisation and International Business* (Oxford: Oxford University Press).

Fishlow, A., Gwin, C., Haggard, S., Rodrik, D. and Wade, R. (1994), *Miracle or Design? Lessons from the East Asian Experience* (Washington, DC: Overseas Development Council).

Hobday, M. G. (1995), *Innovation in East Asia: The Challenge to Japan* (Cheltenham: Edward Elgar).

Kim, K. S. (1994), 'The Korean Miracle (1962–80) Revisited: Myths and realities in strategies and development', in H. Stein (ed.), *Asian Industrialization and Africa* (London: Macmillan), pp. 87–144.

Lall, S. (1991), 'Direct Investment in S.E. Asia by the NIEs: Trends and prospects', *Banca Nazionale del Lavoro Quarterly Review*, 179: 463–80.

Lall, S. (1992), 'Technological Capabilities and the Role of Government in Developing Countries', *Greek Economic Review*, 14(1): 1–36.

Lall, S. (1993), 'Policies for Building Technological Capabilities: Lessons from Asian experience', *Asian Development Review*, 11(3): 72–103.

Lall, S. (1994.a), '*The East Asian Miracle* Study: Does the bell toll for industrial strategy?', *World Development*, 22(4): 645–54.

Lall. S. (1994.b), 'Industrial Policy: The role of government in promoting industrial and technological development', *UNCTAD Review 1994*: 65–89.

Lim, L. (1994), 'Foreign Investment, the State and Industrial Policy in Singapore', in H. Stein (ed.), *Asian Industrialization and Africa* (London: Macmillan), pp. 205–38.

Lipsey, R. G. (1994), 'Markets, Technological Change and Economic Growth', *Pakistan Development Review*, 33(4): 327–52.

Lipsey, R. G. and Carlaw, K. (1995), 'A Structuralist View of Innovation Policy', in P. Howitt (ed.), *Implications of Knowledge-Based Growth for Micro-Economic Policies* (University of Alberta Press).

Moreira, M. M. (1994), *Industrialization, Trade and Market Failures: The Role of Government Intervention in Brazil and the Republic of Korea* (London: Macmillan).

Najmabadi, F. and Lall, S. (1995), *Developing Industrial Technology*, (Washington, DC: World Bank, Operations Evaluation Department).

Selvaratnam, V. (1994), 'Innovations in Higher Education: Singapore at the competitive edge', (Washington, DC: World Bank), Technical Paper No. 222.

OED (1992), *World Bank Support for Industrialization in Korea, India and Indonesia* (Washington, DC: Operations Evaluation Department, World Bank).

Pack, H. and Westphal, L. E. (1986), 'Industrial Strategy and Technological Change: Theory versus reality', *Journal of Development Economics*, 22(1): 87–128.

Rodrik, D. (1994), 'Getting Interventions Right: How South Korea and Taiwan grew rich' (New York: Columbia University and NBER), Paper presented to 20th Panel meeting of *Economic Policy*, October 1994.

Stiglitz, J. E. (1989), 'Markets, Market Failures and Development', *American Economic Review, Papers and Proceedings*, 79(2): 197–202.

UNCTAD (1994), *World Investment Report 1994* (Geneva: United Nations Conference on Trade and Development).

UNCTAD (1995), *World Investment Report 1995, Draft* (Geneva: United Nations Conference on Trade and Development).

Wade, R. (1990), *Governing the Market: Economic Theory and the Role of Government in East Asian Industrialization* (Princeton: Princeton University Press).

Wade, R. (1993), 'Managing Trade: Taiwan and South Korea as challenges to economics and political science', *Comparative Politics*, 25(2): 147–67.

Wells, L. T. and Wint, A. G. (1990), *Marketing a Country: Promotion as a Tool for Attracting Foreign Investment* (Washington, DC: International Finance Corporation).

Westphal, L. E. (1990), 'Industrial Policy in an Export-Propelled Economy: Lessons from South Korea's experience', *Journal of Economic Perspectives*, 4(3): 41–59.

World Bank (1993), *The East Asian Miracle: Economic Growth and Public Policy* (New York: Oxford University Press).

Notes

CHAPTER 1: PARADIGMS OF DEVELOPMENT: THE EAST ASIAN DEBATE ON INDUSTRIAL POLICY

1. I am grateful to Robert Wade and George Peters for comments on an earlier draft.
2. This view of the East Asian experience emerged in writings of Balassa and Krueger, and their associates at the World Bank, which by the late 1980s emerged as the leading proponent of the neoclassical development school.
3. As Lucas (1988) noted at the start of his seminal contribution to 'new' growth theory, the free-trade orientation of the Asian NIEs could not be theoretically explained by their improved resource allocation. An improvement in trade policy could only lead to a once-for-all boost of incomes, it could not raise the rate of growth over the longer term.
4. Of the large literature on this, see Amsden (1989, 1994), Jacobsson (1993), Lall (various), OED (1992), Moreira (1994), Pack and Westphal (1986), Singh (1994), Wade (1990), Westphal (1990).
5. See, among others, Bell and Pavitt (1993), Katz (1987), Lall (1990, 1992, 1993, 1994.b), Pack and Westphal (1987).
6. For a sample of the critiques of the *Miracle* study, see Amsden (1994), Fishlow *et al.* (1994), Kwon (1994), Lall (1994.a), Singh (1994).
7. In developed countries the debate on the role of the government is also very active. See for instance Stiglitz *et al.* (1989).
8. For an interesting analysis of this in Singapore see Selvaratnam (1994). Korea and Taiwan also intervened in the education system to provide the technical skills needed for the industries that were being targeted.
9. World Bank (1993), pp. 90–2.
10. Nelson (1981).
11. The reason is put nicely by John Stuart Mill, a notable and perceptive admission in his vigorous defence of free trade: 'The only case in which, on mere principles of political economy, protecting duties can be defensible, is when they are imposed temporarily (especially in a young and rising nation) in the hopes of naturalising a foreign industry, in itself perfectly suitable to the circumstances of the country. The superiority of one country over another in a branch of production often arises only from having begun it sooner. There may be no inherent advantage on one part, or disadvantage in another, but only a present superiority of acquired skill and experience. . . . *But it cannot be expected that individuals should, at their own risk, or rather to their certain loss, introduce a new manufacture, and bear the burden of carrying on until the producers have been educated to the level of those with whom the processes are traditional.* A protective duty, continued for a reasonable time, might sometimes be the least inconvenient mode in which the nation can tax itself for the support of such an experiment' (J. S. Mill, 1940, p. 922). Emphasis added.

12. Lall (forthcoming).
13. Its 'machinery' exports consist of electronic watches and games rather than capital goods, unlike the other NIEs.
14. *Financial Times*, London, 4 May 1993, 'Survey of Hong Kong', p. 6. Manufacturing employment declined from 45% to 23% of the total in 1980–92, and its contribution to GDP from 27% to 16%.
15. See Lim (1994) on industrial policy, and for a comprehensive analysis of Singapore's selective interventions in education Selvaratnam (1994).
16. For a summary description see Lall (1994.b). For details on Korea see Amsden (1989), Moreira (1994), Westphal (1990), Kim (1994) and Lall and Najmabadi (1995).
17. It enacted a law to promote subcontracting by the *chaebol*, designating parts and components that had to be procured through SMEs and not made in-house. By 1987 about 1200 items were so designated, involving 337 principal firms and some 2200 subcontractors, mainly in the machinery, electrical, electronic and ship-building fields. Generous financial and fiscal support was provided to subcontractors, to support their operations and process and product development.
18. For a comprehensive analysis see Wade (1990). Also see Brautigam (1995) for a concise exposition of Taiwan's industrial policies and the role of selective interventions.
19. For reviews see Chang (1994), Shapiro and Taylor (1990), Streeten (1993).

CHAPTER 2: UNDERSTANDING TECHNOLOGY DEVELOPMENT

1. See Pack (1988).
2. There is a large literature now on the relevance of developing local capabilities to master technologies in developing countries. See Bell *et al.* (1984), Dahlman *et al.* (1987), Enos (1992), Katz (1987), Lall (1987, 1990, 1992), Pack (1992) and Teitel (1984).
3. See Nelson (1987).
4. Stiglitz (1989).
5. The theoretical basis of this approach is mainly the 'evolutionary approach' developed by Nelson and Winter (1982). Also see their earlier (1977) paper on the theory of innovation.
6. On the importance of the institutional aspects of capability acquisition see Chandler (1992) and Enos (1992).
7. Jacobsson (1993) explores the length of the learning period in the engineering industry in Korea.
8. See Lall (1987).
9. See Katz (1984).
10. Nelson and Winter (1982).
11. This approach is succinctly analysed in Nelson and Winter (1977) and Nelson (1981).
12. The World Bank, for instance, adopts this approach in its structural adjustment policies and in its analysis of industrial policy. See Lall and associates (1992).
13. Porter (1990).

14. See Amsden (1989), Lall (1990), Lall and associates (1992), Pack and Westphal (1986), Wade (1990).
15. See Stiglitz (1987).
16. Jacobsson (1993).
17. On the Korean strategy for monitoring infant industries see Pack and Westphal (1986).
18. Lall (1990).
19. This is found for India by Lall (1987).
20. As Rodrik (1992) notes, the links between improved industrial efficiency and trade liberalisation are very tenuous. There are also other strong arguments for a gradualist approach to liberalisation; see Fanelli and Frenkel (1993).
21. Lall (1990).
22. The 1991 *World Development Report* illustrates the World Bank's current stand on the need for 'market-friendly' interventions.
23. See Shapiro and Taylor (1990).

CHAPTER 3: TECHNOLOGY DEVELOPMENT POLICIES: LESSONS FROM ASIA

1. I acknowledge my debt to an unpublished paper on East Asian technology policies by Anne-Marie Carroll, which I have drawn upon at several points.
2. On differences in national technology policy, see Teubal (forthcoming).
3. These types of market failures are analysed in Stoneman (1987).
4. This issue has been addressed recently by endogenous growth theorists like Young (1991).
5. Najmabadi and Lall (1995).
6. Amsden (1989).
7. Dahlman and Sananikone (1990).
8. The government provided NT$20 billion in loans at preferential interest rates for buying equipment, of up to 65% of the investment.
9. By the end of 1992, the government had granted NT$2 billion in matching interest-free loans and NT$1 billion in research grants, mostly to the information and communications industries. The provision of grants was limited to products involving 'high technology', while loans were available, on approval, to most industries.
10. Some of the generalisations that emerge from past experience are as follows. More sophisticated and proprietary technologies, and those involving the valuable brand-names, are more internalised than those involving simpler or older technologies. The more capable the technology buyer, the more it would prefer to (and be able to) externalise the transfer, while the lower the capabilities of the purchaser the more it would rely on internalised transfers, with intensive and prolonged supplier involvement. Large global companies prefer internalised modes of selling technologies, while smaller firms with lower international exposure tend to externalise, and share the cost and risk with the buyer. Technologies aimed at export markets and closely integrated into international production systems would

be more internalised than those aimed at domestic markets or more delinked from global production systems.

11. Gilley (1995).
12. Selvaratnam (1994).
13. Lall *et al.* (1994).
14. This draws on Dahlman and Sananikone (1990). On Taiwan's industrial policies more generally, see Wade (1990).
15. Hobday (1995).
16. Westphal (1990), and Kim (1993).
17. For a study of this strategy in the petrochemicals industry, see Enos and Park (1987).
18. Rhee *et al.* (1984).
19. Export promotion was implemented by general measures like devaluation and general tax incentives, as well as by discretionary measures like access to restricted imports and direct cash subsidies. The state-controlled banking system was used to channel funds into export support, and export performance increasingly became the criterion for credit worthiness. These incentives were backed by powerful direct pressures to export: regular meetings between business leaders and government and detailed targeting of exports at the industry and firm levels (backed by threats of tax auditing and restrictions on imported inputs for poor performers). The export drive also received considerable support from institutional measures like the support of the giant trading and producing conglomerates, assistance to testing and quality-assurance services, export-marketing information, design assistance, and so on. Overt subsidies declined over the 1980s, but institutional support and the indirect influence of the government continued strongly.
20. Lall (1990).
21. Lall (1987).
22. Stiglitz (1989).
23. OTA (1990).
24. Song (1995).
25. Commercial interest rates in the early 1990s have been around 14–15 per cent.
26. Bank Negara Malaysia, *Annual Report 1994*, pp. 175–6.
27. The largest number of approvals was in the electrical and electronics sector (45); followed by the textiles (40); food (39); wood (32); automotive (21); plastics, iron and steel, chemicals and petrochemicals (18 each); machinery and engineering (13); non-metallic minerals (12); and rubber (8).
28. Clifford (1994).
29. *Financial Times*, London, 'Survey of Hong Kong', 4 May 1993, p. 6.
30. Gilley (1995).
31. Soon and Tan (1993).
32. Soon (1994).
33. On the experience of the developed countries see Mowery and Rosenberg (1989).
34. See Chapter 5, note 36, for information on ISO 9000.
35. Information provided by the HKPC.

36. These activities are: information technology, microelectronics, electronic systems, advanced manufacturing technology, materials technology, energy and water resources, environment, biotechnology, food and agrotechnology and medical sciences.
37. For a fuller analysis, see Dahlman and Sananikone (1990), Brautigam (1995), and Hou and Gee (1993).
38. This is the case of the electronics industries in Taiwan and Korea, the former focused more on market niches and the latter more on mass production. See Hobday (1995).

CHAPTER 4: 'THE EAST ASIAN MIRACLE' STUDY: DOES THE BELL TOLL FOR INDUSTRIAL STRATEGY?

1. World Bank (1993).
2. In this view, markets are essentially efficient in developing countries, or, at least, more efficient than governments, and the best development strategy is that of minimal interventions, with no role for 'industrial policy' (selective strategies to promote particular activities).
3. See, in particular, Amsden (1989), Lall (1992), Pack and Westphal (1986), Wade (1990) and Westphal (1990).
4. Interestingly, the *Far Eastern Economic Review* (1992) reports that the publication of this report was opposed strongly by the Bank's management but supported by its executive directors, led by the Japanese. The directors got their way, but the Bank (in contrast to the *Miracle* study) made no subsequent effort to publicise the work.
5. For instance, *The Economist* (1993), one of the leading dogmatists of the free market, which has often portrayed the East Asian economies as purely market driven, worries that the study may give governments the wrong idea.
6. This approach is associated with Balassa, Krueger and Harberger, and also typifies much of World Bank literature and policy work on trade policy.
7. See the references in note 3.
8. First officially propagated in the *World Development Report 1991*.
9. The 'neoclassical' approach to East Asia is not a correct application of neoclassical growth theory. As Lucas (1988) notes, given the standard neoclassical assumption of diminishing returns to investment, liberal trade policies and reallocation according to static comparative advantage only offer a once-for-all benefits, not higher sustained growth.
10. The information considered here applies only to financial markets rather than to technology or product markets.
11. Page 92 of the *Miracle* study (World Bank, 1993).
12. World Bank (1993), p. 84. For a critique see Taylor (1993), and for a more general discussion of the role of the state see Streeten (1993).
13. Lall (1992, 1993).
14. On the case for protection in industrial latecomers John Stuart Mill, perhaps the most powerful advocate of free trade in the history of economic thought, says: 'The only case in which, on mere principles of political

economy, protecting duties can be defensible, is when they are imposed temporarily (especially in a young and rising nation) in the hopes of naturalising a foreign industry, in itself perfectly suitable to the circumstances of the country. The superiority of one country over another in a branch of production often arises only from having begun it sooner. There may be no inherent advantage on one part, or disadvantage in another, but only a present superiority of acquired skill and experience.... But it cannot be expected that individuals should, at their own risk, or rather to their certain loss, introduce a new manufacture, and bear the burden of carrying on until the producers have been educated to the level of those with whom the processes are traditional. A protective duty, continued for a reasonable time, might sometimes be the least inconvenient mode in which the nation can tax itself for the support of such an experiment. But it is essential that the protection should be confined to cases in which there is good ground for assurance that the industry which it fosters will after a time be able to dispense with it; nor should be domestic producers ever be allowed to expect that it will be continued to them beyond the time necessary for a fair trial of what they are capable of accomplishing.' Mill (1940), p. 922.

15. See Jacobsson (1993) on the heavy engineering industry in Korea.

16. On the capital-goods sectors as a 'hub' of technical progress in the USA, see Rosenberg (1986).

17. This is argued by some 'new' growth theorists, who trace differences in sustained growth rates partly to the ability of some countries to specialise in more technology-intensive sets of activities (Young, 1991).

18. See Amsden (1989), Operations Evaluation Department (OED) (1992) and Westphal (1990).

19. As noted above, the failures mentioned include scale economies, interdependence in investment, capital-market deficiencies, strategic negotiations, economies of scope and specialisation, and imperfect appropriability of knowledge.

20. World Bank (1993), p. 301. Thus, the HPAEs mentioned had *more* distorted prices than the import-substituting economies. The deciles are derived from 'an index of outward orientation based on international comparisons of price levels and price variability', which has little to do with normal definitions of trade orientation based on net incentives to sell in different markets.

21. OED (1992).

22. As described in Lall (1992), the level of R&D relevant to industry (financed by productive enterprises) achieved in Korea by the late 1980s, at 1.9 per cent of GDP, was by far the highest in the developing world. It was also higher than that of many developed countries, those that were not technological leaders in industry.

23. The evasiveness is illustrated by the choice of the box (World Bank, 1993, pp. 302–3) following the discussion of openness to technology. The box describes Matsushita's contribution to Malaysian exports, where a more relevant box on market failures and FDI strategy would have described the development of the Japanese automobile industry after foreign investors were thrown out!

24. Ibid., p. 304.
25. For instance, the period 1973–80, which is used to test for Korean structural change, corresponding to the Heavy and Chemical Industry Drive, is certainly too short to gauge the success of the heavy industries launched. Some may need a decade or more to mature to international competitiveness and thus to entry into world markets. See Jacobsson (1993).
26. This would show, for instance, that own exports (as opposed to re-exports) by the least interventionist Dragon, Hong Kong, are now growing at around 3 per cent and manufacturing employment declined by 35 per cent in 1986–92.
27. See Nelson (1981).
28. The tests on Japan, Korea and Taiwan are made for 25 sectors, including such broad categories as metal products, electrical machinery, chemicals, transport equipment and so on.

CHAPTER 5: STRUCTURAL ADJUSTMENT AND AFRICAN INDUSTRY

1. I am grateful to Sunil Mani for his help in collecting the data used in this chapter and to two referees of *World Development* for their helpful comments on an earlier draft.
2. World Bank (1994), p. 1.
3. For a flavour of the recent debates on structural adjustment in Africa, see, among others, Elbadawi (1992), Elbadawi *et al.* (1992), Helleiner (1992, 1994), Killick (1993), Mosley *et al.* (1991, 1993), Stein (1992, 1994.a), Toye (1995). Mosley and Weeks (1994) have a perceptive review of the recent World Bank study.
4. See, for instance, Killick (1993), Lall (1993), Stein (1994.a), and Toye (1995).
5. The first in the series (World Bank, 1993) was on the Asian NIEs and Japan. For a critique see Chapter 4 above.
6. World Bank (1994), p. 61. Also see the article by Husain (1994), the main author of the Bank study, for a discussion of what SAPs involve and how they differ from stabilisation programmes.
7. The principal author of the World Bank's 1994 study, Husain (1994), argues strongly for the need to distinguish between stabilisation and adjustment: 'The term stabilisation is normally used for reducing demand and structural adjustment for stimulating supply. Stabilisation measures mainly focus on financial disequilibrium (i.e. the fiscal and external accounts and the rate of inflation). Structural adjustment policies seek to restructure production capacities in order to increase efficiency and help restore growth. While stabilisation is concerned with the short term, structural adjustment extends over the medium to long term. . . . A reorientation of public expenditures and investment towards human capital and infrastructure, an expansion of exports and efficient import substitutes, efficient trade policies, and liberalisation of prices and deregulation of controls to allow producers to increase output of goods and services are essential characteristics of structural adjustment programmes' (p. 151). He goes on

to argue that 'reform measures must be correctly identified for what they are and what they are not. If, in the first phase of reform, the emphasis was on restructuring the budget and shifting the balance in the current account through expenditure reductions, it would be incorrect to call this a structural reform. *These are stabilisation measures and their impact is very different from that of structural reforms*' (p. 152, emphasis added).

8. See, for instance, World Bank (1991 and 1993).
9. The underlying arguments are analysed in Lall (1994). For reviews of the Bank's *Miracle* study see the April 1994 issue of *World Development*. Singh (1994) has a perceptive and critical discussion of openness and the market-friendly approach.
10. For a perceptive analysis of the assumptions of neoclassical economics on the nature of institutions and markets, and the implications for reform in Africa, see Stein (1994.b).
11. See Mosley *et al.* (1991) and Toye (1995).
12. Toye (1995).
13. See Lall (1992.a and 1992.b).
14. See Lall (1994). On the relevance of East Asia for Africa, see Stein (1994.a).
15. Data from Table 11 of African Development Bank (1994).
16. The policy-improving countries had annual weighted GDP growth rates of 2.7% and 3.4% respectively over 1980–93 and 1990–93 respectively, compared with 1.3% and 0.6% for policy-deteriorating countries and 2.0% and 0.2% for the non-adjusting countries.
17. Data for manufactured exports are taken from the World Bank, *World Development Report 1994*. Figures for exports are missing for some of the countries, especially in the non-adjusting group.
18. World Bank (1994), p. 67; and Leechor (1994).
19. African Development Bank (1994), pp. 57–62.
20. Ibid., Table 27.
21. *African Development Report 1994*, p. 61.
22. Apart from the enclave operation of aluminium processing, or protected activities like government-owned petroleum refining, these include some food processing, furniture, cement, simple metal products, and uniforms for the army or schools.
23. The values of the main non-traditional manufactured exports in 1991 were: aluminium $5.5 m., wood products $6.2 m. (of which furniture accounted for $3.6 m. and other wood products for $2.6 m.), canned foods $0.3 m., tobacco $0.4 m., soaps $0.6 m., machetes and rods $0.8 m., and others $1.3 m.
24. Lall *et al.* (1994). For a shorter exposition, see Chapter 7 below.
25. Ibid.
26. Lall (1992.a).
27. See Lall (1993).
28. World Bank (1994), p. 192, emphasis added.
29. Ibid., p. 196.
30. The World Bank's (1993) *Miracle* study provides many useful insights into how Asian governments improved their intervention capabilities.
31. See, among others, Lall (1992.a) and World Bank (1989).
32. For instance, Korea by itself had 411,000 in technical subjects at the ter-

tiary level in 1990, compared with 111,000 for the whole of sub-Saharan Africa.

33. See Lall *et al.* (1994).
34. The most important development is the International Standards Organisation's ISO 9000 quality-management standards. This is a mandatory requirement for exporting health-related industrial products to the EC, but even where it is not mandatory it is becoming an important competitive asset for all exporters. The advantage of the new ISO series is that it provides an enterprise with a ready set of definitions of what constitutes quality management, and gives clear guidelines on how to set up a quality system. Given the right technical support, therefore, it provides a strong impetus to quality upgrading and technology diffusion – managers have little choice but to meet the standards if they hope to compete in developed-country markets. Its promotion is becoming a major objective of many developing countries, especially in East Asia, but the achievements so far are very uneven. By late 1994, the whole of sub-Saharan Africa, excluding South Africa, had less than ten ISO 9000 certificates in 1993, while Singapore by itself had over 550.
35. On the Hong Kong Productivity Council (HKPC) see Chapter 3 above.

CHAPTER 6: MALAYSIA: INDUSTRIAL SUCCESS AND THE ROLE OF GOVERNMENT

1. Data are taken from Lall *et al.* (1994). Also see Government of Malaysia (1991).
2. The use of US data rather than Malaysian is deliberate. A skill classification based on US data is likely to reflect better the structural characteristics of industry and is more useful to compare the composition of exports across countries.
3. Lall *et al.* (1994).
4. See studies and estimates of sources of manufacturing growth cited in Salleh and Meyanathan (1993).
5. See Ali (1992), and various papers in Jomo (1993).
6. Data cited in Salleh and Meyanathan (1993).
7. For a review of the debate see Lall (1994.a). Also see Chang (1994), Colclough and Manor (1992), Moreira (1994), UNCTAD (1994) and Wade (1990).
8. World Bank (1993).
9. See, among others, Amsden (1994), Kwon (1994), Lall (1994.b), Rodrik (1994), and Wade (1994).
10. Lall (1992).
11. Lall (1994.b).
12. This section draws on Salleh and Meyanathan (1993) and Brown (1993).
13. Interviews with MIDA (Malaysian Industrial Development Authority).
14. Salleh and Meyanathan (1993), p. 14.
15. On Korean interventions see Amsden (1989), Lall (1994.a) and Moreira (1994).
16. Jacobsson (1993).

17. Lall *et al.* (1994).
18. The technology plan targeted the following technologies for future development: automated manufacturing, microelectronics, advanced materials, biotechnology and information technology. A fund of M$600 million was established to support research in these areas. For details see Lall *et al.* (1994).
19. Brown (1993), pp. 44–5.
20. World Bank (1993), pp. 310–11.
21. See Ali (1992), Government of Malaysia (1991), Jomo (1993), Lall *et al.* (1994).
22. Data are not collected in Malaysia directly on the origins of exports. However, there is sufficient evidence at the sectoral level for all major export industries to conclude that the bulk of exports comes from MNCs.
23. The government has set a target for Malaysia to become a fully developed economy by 2020. This involves ambitious rates of growth of manufactured exports and industrial production.
24. For instance, data on FDI by the Malaysian Industrial Development Authority (MIDA) show that approvals in electrical and electronics peaked at $1.5 billion in 1990 and were at half that figure ($0.7 billion) in 1993.
25. See Brown (1993).
26. See Wade (1994).

CHAPTER 7: SKILLS AND CAPABILITIES IN GHANA'S COMPETITIVENESS

1. See Fontaine (1992), Leechor (1994), Steel and Webster (1992) and World Bank (1994).
2. For a fuller analysis see Lall *et al.* (1994).
3. See, for instance, Fagerberg (1988) and Lazonick (1993).
4. See Bell and Pavitt (1993), Enos (1992), Katz (1987), Lall (1987, 1992.a, 1993.a and b), and Wignaraja (forthcoming).
5. See, for instance, World Bank (1991) and (1993).
6. The underlying arguments are analysed in Lall (1994).
7. See Mosely *et al.* (1991) and Stein (ed.) (1995).
8. For a recent attempt by the World Bank, on Africa, see its 1994 study on *Adjustment in Africa*.
9. World Bank (1994), p. 67, and Leechor (1994).
10. African Development Bank (1994), pp. 57–62.
11. Ibid., Table 27.
12. Ibid., p. 61.
13. Apart from the enclave operation of aluminium processing or protected activities like government-owned petroleum refining, these include some food processing, furniture, cement, simple metal products, and uniforms for the army or schools.
14. The values of the main non-traditional manufactured exports in 1991 were: aluminium $5.5 m., wood products $6.2 m. (of which furniture accounted for $3.6 m. and other wood products for $2.6 m.), canned foods $0.3 m.,

tobacco $0.4 m., soaps $0.6 m., machetes and iron rods $0.8 m., and others $1.3 m.

15. Lall *et al.* (1994).

16. See Lall (1992.b) for an explanation of what these categories comprise.

17. Market segmentation may well exist in Ghana (as in all developing countries, formal credit markets tend to be biased against small firms). The analysis of the panel data in Lall *et al.* (1994) suggests that such segmentation exists, but that it does not account for the whole difference in performance between firms of similar sizes. The other factors that show up as important determinants of competence in the next section suggest that segmented factor markets play a relatively minor role.

18. In the case of large professionally managed firms (in this sample mainly multinationals) the entrepreneur is taken to be the chief executive officer of the firm.

19. Enterprises of larger size and with more mature organisational structures do not need to rely on the (partly random) presence of a technologically gifted person to catalyse the firm. They would tend to have institutional mechanisms to identify, recruit and assign due responsibility to such persons. This is one of the advantages of large and functionally specialised organisations that was mentioned in the analytical framework.

20. For a fuller analysis of the acquisition of technological capabilities in the sample, see Wignaraja (forthcoming).

21. The Kumasi Technical Institute trains young people in wood working, but does not seem to provide more advanced employee training.

22. Lall (1992.a).

23. See Lucas (1994).

24. See Stein (1995) for descriptions of industrial policy in various NIEs and Japan. The NIEs and Japan provided substantial technical extension services to industry in the areas of quality control and metrology, research and development, information on sources of technology and assistance in the purchase of foreign technology.

25. The World Bank's (1993) *Miracle* study provides many useful insights into how Asian governments improved their intervention capabilities.

CHAPTER 8: FOREIGN DIRECT INVESTMENT POLICIES IN THE ASIAN NIEs

1. This abstracts from traditional concerns about transfer pricing, bargaining, predatory conduct and so on.

2. The World Bank's 1993 *Miracle* study claims that good macro-management, export orientation and 'market-friendly' interventions to strengthen human capital are sufficient to explain East Asian industrial success; but this fails to take into account the very marked differences in industrial-policy objectives, instruments and achievements in the region (Lall, 1994.a).

3. *Financial Times*, London, 4 May 1993, 'Survey of Hong Kong', p. 6. Manufacturing employment declined from 45% to 23% of the total in 1980–92, and its contribution to GDP from 27% to 16%.

4. *Far Eastern Economic Review*, 26 May 1994, p. 69.

5. See Lim (1994) on industrial policy, and for a comprehensive analysis of Singapore's selective interventions in education see Selvaratnam (1994).
6. Hobday (1995).
7. Singapore established the National Technology Board (NTB) to attract functional headquarters of MNCs for research and development. The NTB will direct the expansion of a research and development infrastructure for new industries, such as agro-technology, biotechnology, robotics and automation. Singapore also established several government-support research centres, including the Singapore Science Park, the Institute for Molecular and Cell Biology, the Institute of Systems Science, and the Information Technology Institute. A new university devoted to science and technology will double Singapore's R&D expenditure to over half a billion dollars. Singapore's Technology Development Centre (TDC) helps local companies identify their technology requirements and design appropriate strategies for upgrading their operations. Since its establishment in 1989, the centre has sent its multi-disciplinary staff of consultants and engineers on over 300 plant visits, and provided more than 130 companies with various forms of assistance including sourcing of foreign experts and equipment, and advice on process improvement and product development.
8. For a summary description see Lall (1994.b). For details on Korea see Amsden (1989), Kim (1984), Moreira (1994), Najmabadi and Lall (1995) and Westphal (1990).
9. Korean strategy in technology development in electronics is analysed by Hobday (1995).
10. On semiconductors alone, four leading South Korean *chaebol* spent Won 1.8 trillion on capital investment and Won 300 billion on research and development in 1989–90. Their size and financial resources give them a clear advantage over similar companies in Taiwan, Hong Kong, and Singapore.
11. For a comprehensive analysis see Wade (1990). Also see Brautigam (1995) for a concise exposition of Taiwan's industrial policies and the role of selective interventions.
12. Brautigam (1995), p. 171.
13. In an attempt to acquire semiconductor design and production capability, in 1974) the Taiwanese government formed the Electronic Research and Service Organization, authorizing ERSO to recruit a foreign partner to help develop and commercialise the technology. In 1976, ERSO opened the country's first model shop for water fabrication, and a year later signed a technology-transfer agreement with RCA in integrated circuit design (Wade, 1990, pp. 103–4).
14. There was, nevertheless, a strong political commitment to promoting local capabilities. There are other large economies with sizeable industrial sectors, such as Mexico, that have chosen to remain highly dependent on imported technologies. As a consequence, R&D by enterprises in Mexico is around 0.02% of GDP as compared with 1.8% in Korea, when both have roughly equal values of manufacturing value added (Najmabadi and Lall, 1995).
15. See Wade (1990, 1993).

Index

Note: page numbers in *italics* denote tables or figures.